Introductory Sports Engineering and Technology

Introductory Sports Engineering and Technology provides a fundamental primer to the world of Sports Engineering and Technology, including the key themes that dominate design and practice. The text has been created to provide a comprehensive overview of key principles and is developed from established and recognised education providers.

This new text includes case studies and worked examples to facilitate deeper learning and prepare readers for a high-performance future in Sports Engineering and Technology. Sports performance is a complex and increasingly technology-enhanced environment, incorporating a broad range of disciplines and topics. The sports engineer or technologist plays a key role within this environment, applying fundamental techniques derived from maths and physics to solve problems, analyse opportunities, and optimise equipment and practice to maximise performance outcomes and reduce injury risk.

This book is the first to offer a concise overview of the fundamental concepts required for budding sports engineers and technologists and is key reading for undergraduate and master's students of Sports Engineering and Technology degree courses, as well as undergraduate engineering, technology, and product design students.

Tom Allen, PhD, is a Reader in the Department of Engineering at Manchester Metropolitan University, UK. He is a Fellow of the Institution of Mechanical Engineers (IMechE) and serves as Editor-in-Chief of *Sports Engineering*.

Nick Emerson, PhD, is a Senior Lecturer and Associate Head of the School of Product Design within the Faculty of Engineering at the University of Canterbury, New Zealand. He is a Chartered Mechanical Engineer in the UK and Australia and a member of the Design Institute of New Zealand.

John Hart, PhD, is a Senior Research Fellow in the Sports Engineering Research Group at the Advanced Wellbeing Research Centre, Sheffield Hallam University, UK. He leads the MSc programme in Sports Engineering and Artificial Intelligence at Sheffield Hallam University.

Stefan Schwanitz, PhD, is a Senior Lecturer in Sports Engineering at Chemnitz University of Technology, Germany. He has taught on both the bachelor's and master's degree programmes in Sports Engineering at Chemnitz for over 15 years.

Introductory Sports Engineering and Technology

Tom Allen, Nick Emerson, John Hart, and Stefan Schwanitz

Routledge
Taylor & Francis Group
NEW YORK AND LONDON

Designed cover image: Getty images

First published 2026
by Routledge
605 Third Avenue, New York, NY 10158

and by Routledge
4 Park Square, Milton Park, Abingdon, Oxon, OX14 4RN

Routledge is an imprint of the Taylor & Francis Group, an informa business

ISBN: 978-1-032-53251-6 (hbk)
ISBN: 978-1-032-53250-9 (pbk)
ISBN: 978-1-003-41108-6 (ebk)

DOI: 10.4324/9781003411086

Typeset in Sabon
by Apex CoVantage, LLC

Contents

Figures

Tables

Foreword

Sports Engineering is a discipline that continues to grow in relevance, complexity, and opportunity. It brings together the rigour of engineering with the passion and dynamism of sport, offering a unique platform for innovation, impact, and societal benefit. As sport evolves, driven by advances in materials, data, manufacturing, and digital technologies, so too must the education that underpins it. This textbook is a timely and important contribution to that evolution.

Education in sports engineering plays a vital role in preparing the next generation of engineers to meet the challenges and possibilities of modern sport. It equips students with the technical knowledge to design safer helmets, more efficient footwear, smarter wearables, and more inclusive sporting environments. But it also fosters a mindset of curiosity, responsibility, and collaboration – qualities essential for working across disciplines and engaging with athletes, governing bodies, and industry.

This book offers a comprehensive and structured introduction to the field. It begins by establishing the foundations of sports engineering, including its history, societal role, and career pathways. It then explores the full development process of sporting goods, from design through to materials selection, manufacturing, testing, and evaluation. Each chapter is grounded in real-world examples and supported by case studies that bring the theory to life, such as the co-design of climbing wearables, the evolution of ski boot safety, and the use of additive manufacturing in elite cycling.

Importantly, the book does not stop at the fundamentals. It looks ahead to the future of sport and technology, covering emerging topics such as smart equipment, mass market customisation, artificial intelligence, and sustainability. It also addresses the ethical and regulatory challenges that come with innovation, helping students to think critically about the broader impact of their work. The discussion of super shoes, the role of big data in performance analysis, and the implications of reshoring manufacturing all reflect the authors' commitment to preparing students for the realities of a fast-moving and socially responsible industry.

Written by a team of experienced educators and researchers, this textbook reflects the depth and diversity of the discipline. It is rigorous, readable, and relevant. It is designed to support learning, spark innovation, and lend credibility to a growing and vital field. As someone who has worked in this space for many years, I believe this book will become a cornerstone for students and educators alike.

Steph Forrester

Steph Forrester is Director of the Sports Technology Institute at Loughborough University. Her career spans elite sport, academia, and industry, with contributions to artificial turf, footwear, and performance monitoring. She has secured major funding and shaped international standards through collaborations with many sports brands and international governing bodies.

Acknowledgements

Thanks to Dr Tasneem Sabir for input on the textiles sub-section of Chapter 6.

Thanks to Dr Nathan Elliott, Yudai Washida, Dr David Cole, and Dr Marcos David Moya Bencomo for kindly providing the career case studies.

1 An introduction to Sports Engineering and Technology

1.1 A brief history of the discipline of Sports Engineering

This opening chapter introduces the discipline of Sports Engineering and Technology and outlines its historical development and societal role. It also summarises typical job roles of sports engineers. Throughout this textbook, the term "Sports Engineering" is used to capture the transdisciplinary field that integrates principles from mechanical and electrical engineering, physics, materials science, and biomechanics to solve problems in sport and exercise. Sports Engineering has a rich and varied history, and the field has become highly organised and structured in recent years. Today, the field has dedicated societies, conference series, academic journals, and education programmes. A valuable resource for learning more about the history of Sports Engineering is the pop-science book *Advantage Play: Technologies that Changed Sporting History* [1].

Within this textbook, the term "sporting goods" refers to any product used for sport and exercise, including sports equipment, clothing, footwear, and wearable sensors. The development of sporting goods also has a rich and varied history. While early innovations in sporting goods were driven by artisans and inventors, the industrial revolution enabled the mass production of items such as bicycles and skis. Some iconic sporting goods companies were established during this period. For example, the tannery that later became Mitre Sports International Ltd, or simply Mitre, was opened by Benjamin Crook in 1817 in Huddersfield, England.

A well-known chapter in sporting goods history is the story of the Dassler brothers. In 1919, Rudolf "Rudi" and Adolf "Adi" Dassler co-founded "Gebrüder Dassler Schuhfabrik", or Geda, in Herzogenaurach, Germany. About 30 years later the brothers split and formed their own companies, the now globally recognised brands Puma (Rudi) and adidas (Adi). adidas famously capitalised on Germany's win at the 1954 Fédération Internationale de Football Association (FIFA) World Cup – referred to as the "Miracle of Bern" – by supplying football boots with screw-in studs for better grip in wet conditions. Although similar patents existed for screw-in studs (e.g., [2–7]), the brand leveraged the event to build its identity and influence. This example highlights how the interplay between product innovation, marketing, and global events can influence the success of sporting goods brands.

While adidas exemplifies how privately owned companies used sport as a platform for brand growth, the 1960s also saw governments recognising the strategic value of elite sport. In 1963, the Institute for Research and Development of Sports Equipment (*Forschungs- und Entwicklungsstelle für Sportgeräte*, FES) was established in East Berlin. As one of the earliest institutional examples of Sports Engineering, the FES was founded with the goal of providing athletes of the German Democratic Republic (GDR) with

DOI: 10.4324/9781003411086-1

technically superior equipment. Olympic and World Championship victories were regarded not only as sporting success but also as proof of the superiority of the socialist system. Consequently, the GDR invested significant state resources into high-performance sport, using it as a tool for international prestige and ideological assertion. The FES played a central role in this strategy, contributing to numerous medal-winning performances in disciplines such as rowing, track cycling, and bobsledding.

As innovation advanced, the need for structured evaluation of equipment performance and safety became apparent. In 1973, the National Operating Committee on Standards for Athletic Equipment (NOCSAE) was founded in the United States. Their helmet testing standard (ND002 [8]) introduced the use of biofidelic headforms and impact severity metrics, laying the foundation for safety standards in sports. Standards for safety devices remains a key topic for many sports engineers, particularly those concerning helmets and head protection.

While these developments marked the beginning of more formal regulation in sport, they were preceded by a long tradition of scientific interest in sporting phenomena. Long before performance standards or computational simulations were introduced, scholars and natural philosophers were exploring the physical principles underlying sport through observation, measurement, and theory. This curiosity laid the intellectual foundations upon which modern Sports Engineering would later be built. One of the earliest documented examples is from 1877, when Lord Rayleigh described the "irregular flight of a tennis ball" [9], marking an early link between physics and sport.

The 20th century saw steady growth in sport-related science, with increasing interest in the aerodynamics of balls and projectiles, and in biomechanics. By the 1960s, researchers used high-speed video and force measurement tools to explore athlete performance. The 1980s and 1990s marked a technological shift. During this period sensors such as strain gauges and accelerometers enabled more detailed measurements. Elsewhere advances in computing facilitated the use of simulation tools like finite element analysis (FEA) and computational fluid dynamics (CFD) (see Part 1 of ref. [10]). Sports brands increasingly collaborated with academics to refine their products. This era also saw the emergence of Sports Engineering as a distinct academic discipline.

The first *Engineering of Sport Conference* was held in Sheffield in 1996 [11]. In 1998, the International Sports Engineering Association (ISEA)[1] was founded, alongside its journal *Sports Engineering*.[2] The quote that opened the first issue of *Sports Engineering* remains relevant: "Sports engineering is about the design of equipment, surfaces, apparel or technology which enhances the sport to which it is applied" (Prof. Steve Haake [12]).

Within the Institution of Mechanical Engineers (IMechE), Sports Engineering was formally recognised in 2008 with the launch of *Proceedings of the IMechE, Part P: Journal of Sports Engineering and Technology*.[3] The size of the sporting goods industry was the focus of the first editorial [13], and the following extract should be welcome news to students wishing to enter the sector: "The world's largest sporting goods brands enjoy multi-billion pound turnovers and employ tens of thousands of people world-wide" (Professor Mike Caine [13]).

The IMechE has also published several influential reports, such as "Sports Engineering: an unfair advantage?" (2012) [14], and "Sustainable, inclusive, innovative: the role of engineering in sport" (2023) [15], and most recently, "Engineering better sports venues" (2024) [16].

The annual Japanese symposium on *Sports Engineering and Human Dynamics* was first held in 1986 [17]. It is considered the longest-standing conference series on sports engineering. In 2015, Sports Engineering and Human Dynamics became a technical division of the Japan Society of Mechanical Engineers (JSME) [17]. Today, Sports Engineering is supported by an international community, comprising research institutions, universities, professional societies, and industry. Students are encouraged to explore the ISEA website[1] and other organisations to learn more about the academic and professional networks within the field.

1.2 The role of Sports Engineering in society

Disciplines that make up Sports Engineering include, but are not limited to, aspects of design, materials, mechanical and electrical engineering, as well as sports physics, sports science, and sports medicine, particularly biomechanics (Figure 1.1a). Sports engineers often work in an interdisciplinary manner across various areas of application. These areas can include athlete performance enhancement and injury risk management, developing products, regulating and officiating competitions, improving the spectator experience, and increasing accessibility and uptake of sport and physical activity, including for people with physical and mental impairments (Figure 1.1b). Indeed, Sports Engineering can help to progress adaptive sport, by removing barriers to participation, enhancing performance, and reducing injury risk [18, 19]. With the emergence and popularity of activities like snowboarding, mountain biking, skateboarding, and jogging, Sports Engineering is making important contributions to recreation and health, as well as competition.

The rapid innovation cycles in the sporting goods industry provides an ideal test arena for new technologies and processes [15, 20, 21], such as advanced materials, wearable sensors, and smart manufacturing and digitalisation. This means Sports Engineering can deliver broad societal benefits. One example is the early uptake of metamaterials in sporting goods [22–24], including helmets [25, 26], body protection [24], and footwear, helping to pave the way for their use in other sectors. Metamaterials are engineered materials with unique, unusual, and enhanced properties, which cannot be achieved with the constituent materials alone [22, 27] (see Chapter 6 for more detail). Other examples include the development of wearable sensors for exercise monitoring and encouragement, and trialling of more environmentally friendly manufacturing processes on mass-produced sporting goods [20].

1.3 The impact of technology on sport

The development of sporting goods has a rich and interesting history (see examples in Figure 1.2). This development has been driven by innovations in design, materials, and manufacturing, coupled with advances in engineering analysis tools and techniques. For many sports, natural materials, like leather and wood, have been superseded by those that are synthetic, like plastics. Examples of innovations in sporting goods include the use of vulcanised rubber in sports balls and shoes, the introduction of laminated skis, and the use of fibre-matrix (reinforced) composites (e.g., fibreglass and carbon fibre) to make "over-sized" tennis rackets [1, 28–36]. Fibre-matrix composites were developed for aerospace applications during World War Two, and due to their high stiffness-to-weight ratio (specific stiffness – see Chapter 5) were soon applied to sporting goods. Early examples of the use of such composites in sport include surfboards, golf club shafts, and skis, followed by tennis rackets, snowboards, and ice hockey sticks (see Part 2 of ref. [10]).

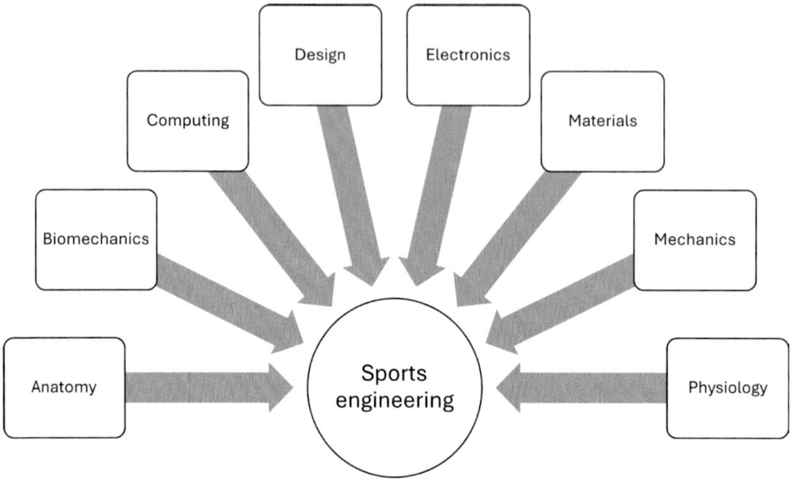

(a)

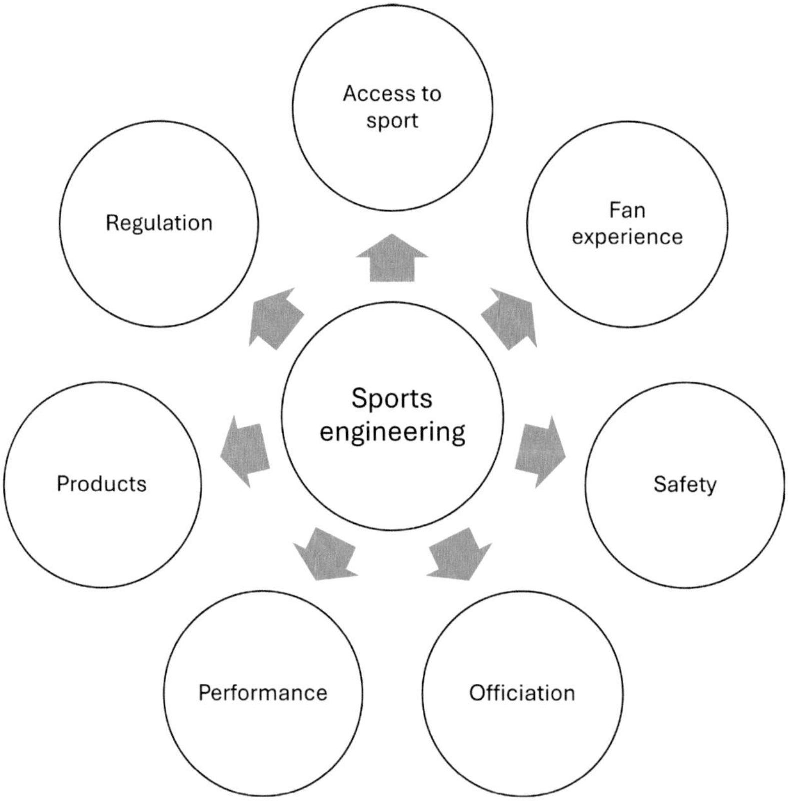

(b)

Figure 1.1 Diagrams showing examples of (a) topics feeding into Sports Engineering and (b) application areas stemming from Sports Engineering.

Figure 1.2 Photos showing examples of early (left) and modern (right) sporting goods: (a and b) football boot, (c and d) ice skate, (e and f) golf club, (g and h) tennis racket, (i and j) football, (k and l) skateboard, (m and n) ski, boot, and binding.

Figure 1.2 (Continued)

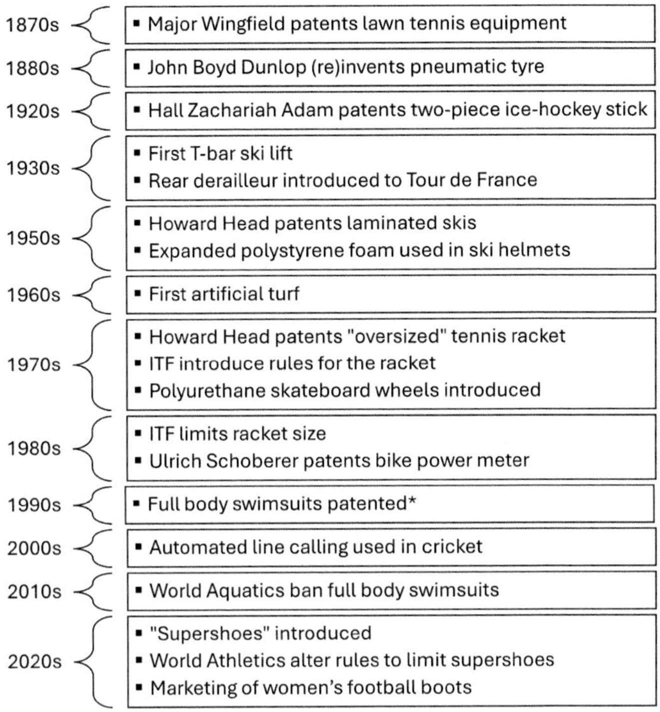

Figure 1.3 Diagram listing examples of historical developments in Sports Engineering and Technology [31, 32, 37–41]. Note these are just indicative and are not meant to represent all the important developments.

* Debuted at 2000 Sydney Olympics

The end of the 20th century saw a global shift in where sporting goods were made. Brands moved production from the West to countries in the Far East with lower labour costs. There are recent examples of brands using more automated factories to bring some production back to the West (e.g., "reshoring"), using Industry 4.0 principles such as smart manufacturing and digitalisation. Elsewhere, the miniaturisation of electronics has led to the emergence of wearable sensors and "smart" equipment and clothing [18]. There is now widespread uptake of wearable technologies, including "fitness trackers". The timeline in Figure 1.3 highlights examples of sporting goods innovations and trends.

There are many historical examples of how equipment developments have influenced sport and sometimes led to sporting governing bodies changing their rules [1, 42]. These examples include:

- The contribution of developments in:
  - Bikes to the hour record [43–47]
  - Golf clubs to longer drives [48, 49]
  - Swimsuits to swimming performance [50]

- The "Spaghetti string" tennis racket, which allowed the wielder to impart high spin rates on the ball [51, 52]
- The recent (circa 2016) introduction of lightweight distance running shoes with thick compliant foam midsoles containing stiff curved plates (i.e., "super shoes") [53–56]

The 1986 javelin rule update to change flight characteristics is also noteworthy [45]. Examples like these are detailed in Chapter 9. Two others are now briefly summarised to set the scene for the textbook. These two examples are the effect of:

i. High shank ski boots on lower limb injuries
ii. Pole vault material on vault height

1.3.1 *High shank ski boots and lower limb injuries*

Injury epidemiology is the study of the distribution and causes of injuries. By collating historical data on the distribution of injuries for a given sport, sports engineers can gauge the potential causes, including those linked to equipment, facilities, and venues. This approach is often applied to winter sports, such as alpine skiing and snowboarding, including by the International Society for Snowsports Safety (ISSS).[4] Such research has found that snowboarders are more prone to upper limb injuries than skiers, who are more likely to injure their lower limbs [57, 58]. An example will now be used to show how epidemiological data can provide insight into the effect of equipment design on injury risk in skiing.

Figure 1.4 shows historical data on the proportion of knee and below knee injuries amongst skiers from the 1940s to 1990s [59]. Below knee injuries include tibia and ankle fractures and ankle sprains. A dashed vertical line approximates the start of the transition to higher and stiffer boots, leading to the modern high shank ski boot (see Figure 1.2m and n). The data indicates that in the 1940s and 1950s, when skiers used soft boots and non-release bindings, the proportion of below knee injuries was almost constant at around 15%. During this period the proportion of knee injuries was also relatively constant, at around 20%. As ski boots became higher and stiffer, the two datasets diverge, with the respective proportion of knee and below knee injuries increasing and decreasing. Unlike knee injuries, tibia and ankle injuries from skiing are now rare.

The results in Figure 1.4 suggest that high shank knee boots protect the lower leg. While these boots appear to drastically reduce the chance of a tibia or ankle injury, they also appear to increase the risk of a knee injury, probably by transferring loads up the leg. Such knowledge can help sports engineers to develop safer equipment, and associated standards and guidelines that enhance safety. In this case, sports engineers continue to focus on improving the design of ski boots and bindings, and associated standards (e.g., ISO 5355:2019 [60], ISO 9462:2023 [61]), to reduce knee injury risk (e.g., bindings that release during a fall).

1.3.2 *Pole vault material and vault height*

Historical data can also offer insight into how equipment may affect sporting performance. For example, Figure 1.5 shows how the male pole vault world record has progressed since the 1840s. The dashed vertical lines mark the first record with each new pole material. The data suggests that pole material influences performance. A clear increase in world record heights followed the introduction of bamboo poles around the 1900s and then again with

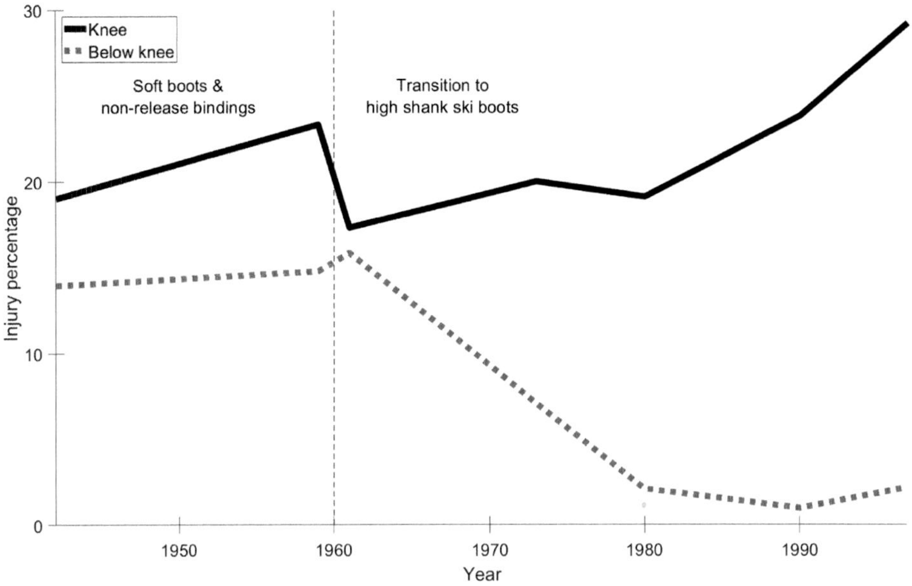

Figure 1.4 Graph showing historical changes in knee and below knee injuries as a percentage of total ski injuries from 1942 to 1997. Below knee injuries include tibia and ankle fractures and ankle sprains. The beginning of the transition to high shank ski boots in c1960 is marked by a dashed vertical line. Data reproduced from [59]. Note the data presented here only goes to the 1990s, and including data after this period would help sports engineers to gauge the effect of more recent technological developments, such as improvements in ski bindings.

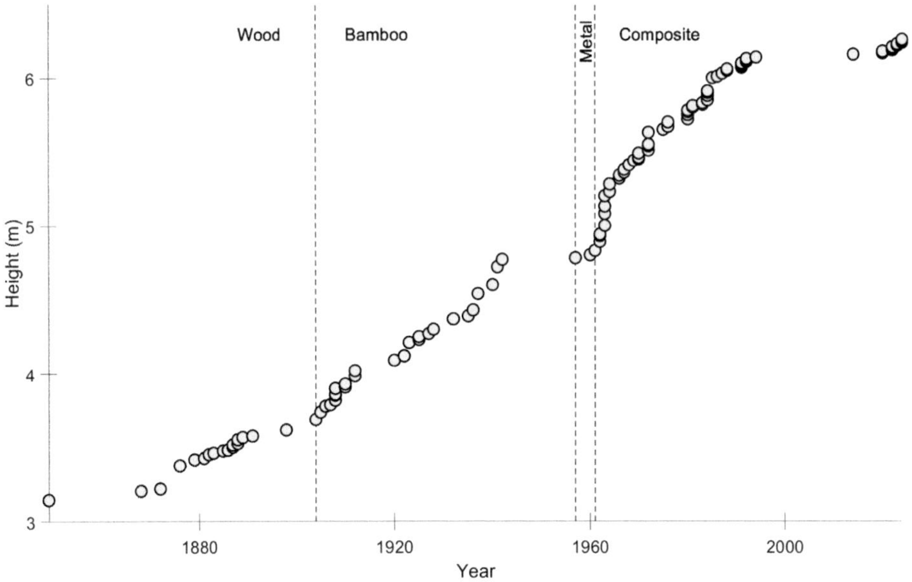

Figure 1.5 Graph showing the historical progression in male pole vault world record height from 1849 (first amateur record) to 2024. The vertical lines mark the first record with a new material, in the order of bamboo (1904), metal (1957), and fibre-matrix composites (1961). Before bamboo, wooden poles were made of ash, oak, spruce, cedar, or hickory. Data taken from Wikipedia and checked against ref. [62].

the move to fibre-matrix composites around the 1960s. These increases in the world record height were obviously not only due to technological advances. Haake [45] estimated that about a third of the increase was due to the pole material. Indeed, such historical data only gives an indication of how a technological development may have influenced performance or injury risk. Sports engineers can use various techniques, such as calculations, simulations, and laboratory and field tests [18], to collect more evidence and better gauge any effect of a new technology. These techniques and topics will be covered within the textbook.

Sports engineers must also consider the broader implications, including those related to ethics, inclusivity, and fairness, of using engineering and technology to enhance sporting performance [42, 63, 64]. Some athletes may not have access to the latest equipment. This equipment may be outside their budget, it may not be commercially available, or their sponsor may simply not offer it. Likewise, not all athletes will respond and benefit from "better" equipment in the same way. Sports engineers should consider how these factors could influence the data they are collecting and working with, as well as sporting world records, to avoid bias and forming incorrect conclusions. Some equipment may improve performance but decrease safety (see Figure 1.4) or have a higher environmental impact. This could be due to its lower durability or use of less sustainable materials and manufacturing processes.

1.4 Working in Sports Engineering

The Sports Engineering industry is broad and diverse, with various options available (see Figure 1.6 and Table 1.1). The following list highlights some organisation types where sports engineers work:

- *Officiating systems providers*, including those developing tools to assist the referee, detect goals, or determine whether a ball lands within a line boundary
- *Organisations focused on safety*, such as the BFU (Swiss council for accident prevention)
- *Organisations supporting athletes and teams*, such as the UK Sports Institute, the Australian Institute of Sport, and the FES
- *Sporting governing bodies*, such as national or international Olympic committees, and specific sports federations (for cycling, track and field, martial arts, team sports, etc.)
- *Sporting goods brands*, ranging from start-ups to multinational corporations
- *Test houses*
    - Including those that certify:
        - Equipment like helmets and body protection inhouse
        - Facilities like courts and pitches
        - Officiating systems
    - Some sporting governing bodies do certification testing themselves, whereas others outsource such work to external test houses.
- *Universities and research institutes*, including education, research, enterprise, and public engagement activities. These universities often work with the organisation types listed here, providing access to the latest knowledge, methods, and developments across science and engineering
- *Wearable sensor and tracking system providers*, including those developing commercial and consumer products for gauging sporting performance, assessing injury risk, and monitoring physical activity, as well as systems for research purposes

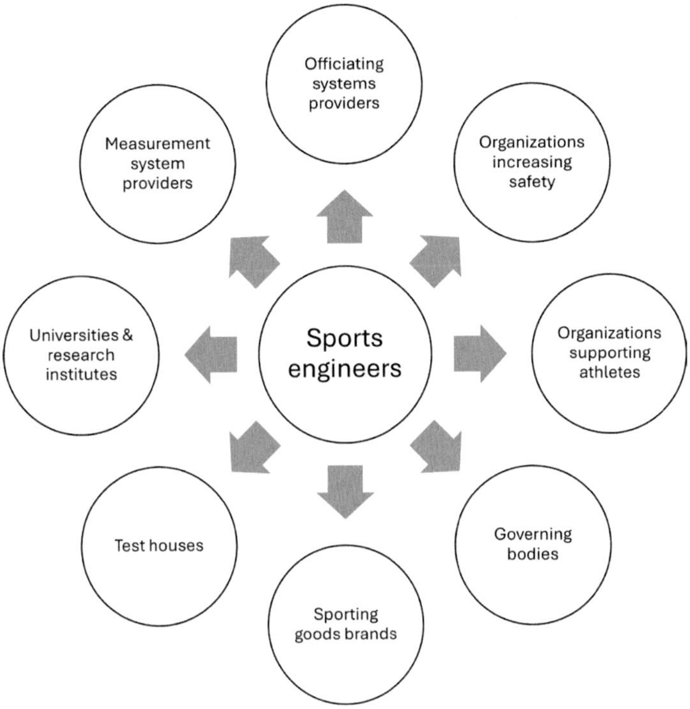

Figure 1.6 Diagram showing examples of the organisation types where sports engineers work.

The modern sports engineer must understand the basic physics of the system they are working with, as well as the requirements of the end user. This user could be an elite athlete, a competent amateur, or a beginner. End users can also vary in age, confidence, physique, and fitness, as well as experience and skill level. The core technical knowledge and skills required to work in the Sports Engineering sector include engineering, physics, design, development, materials, manufacturing, and biomechanics, to name a few. Sports engineers must also communicate effectively, often while solving multifaceted problems within a transdisciplinary team. People working in Sports Engineering traditionally have a background in an established subject such as physics, sports science, or engineering, typically mechanical, materials, design, or electrical engineering. While studying these subjects remains a viable route into Sports Engineering, graduates can now enter the field via dedicated education programmes at undergraduate level or specialise through further postgraduate study.

The sporting goods industry can be influenced by global events and trends, such as changes in political regimes, the health of the economy, major incidents, or a pandemic. For example, during the Covid-19 pandemic, more people engaged in outdoor sports like golf and cycling. An increased demand for associated sporting goods coincided with global supply chain issues. These issues were caused by a combination of reduced productivity due to enforced social distancing and the Ever Given tanker blocking the Suez Canal. This means global events and trends can influence the role of sports engineers within the sporting goods industry.

Table 1.1 Employer case studies

Dr Nathan Elliott

Undergraduate degree	BSc Sports and Materials Science, the University of Birmingham, 2009
Postgraduate degree	MSc Sports Engineering, Sheffield Hallam University, 2010
Research degree	PhD Sports Engineering, Sheffield Hallam University, 2015
Role and employer	R&D Manager and Innovation Team Lead, HEAD Sports GmbH
What I do	My current role at HEAD involves managing all technical developments in the 'non-string bed' team. This includes developing and researching new high-performance technologies and material options for pickleball and padel equipment. Both sports are fast-growing and dynamic, which offers scope for trying and testing new product ideas and innovations. I regularly present and explain HEAD's new product technologies and innovations at global sales meetings and in videos on social media channels. I also get the opportunity to travel and build excellent working relationships and networks with colleagues in Europe, USA and Asia. I also lead the racket sports innovation team whose purpose is to instil curiosity, positive examples, facilitate ideation, foster continuous improvement, and challenge the status quo across the wider rackets sports division.
Academic interests	Sports equipment performance development and testing
Favourite sports	Running, cycling, tennis, pickleball, skiing
Advice for students	Remain curious, foster continuous learning and improvement, and look "outside the box" for inspiration. Build your network by taking opportunities and engaging in sports engineering events and conferences.

Yudai Washida

Undergraduate degree	BEng Mechanical Engineering, Tohoku University, 2012
Postgraduate degree	MSc Sports Engineering, Sheffield Hallam University, 2013
Role and employer	Chief Engineer, Advanced Development Section, Global Equipment Product Department, Mizuno Corporation

(*Continued*)

Table 1.1 (Continued)

What I do	In our section, we are committed to research and development in design techniques and digital solutions for the development of equipment and services for competitive sports such as baseball bats, golf clubs/balls, and badminton rackets. My role in the section involves constructing measurement technology for assessing player preferences as well as building an automated design system to incorporate these parameters into the design. The interesting aspect of building an automated design system lies in contemplating deeply about how to quantitatively evaluate sensations and incorporate them into design parameters. As we are a manufacturer handling equipment for various sports, we are undertaking initiatives to utilise our knowledge cross-sectionally.
Past employers	Research Assistant, University of Tsukuba
Academic interests	Parametric modelling, finite element analysis, optimisation
Favourite sports	Tennis, baseball
Advice for students	In designing sports equipment, the foundations I acquired in Sports Engineering bring a synergy with the new technologies I am learning daily. Furthermore, my ongoing interactions with classmates from all over the world also serve as motivation to maintain my career in the long run.

Dr David Cole

Undergraduate degree	BSc Sports Technology, Loughborough University, 2015
Postgraduate degree	PhD Mechanical/Sports Engineering, Loughborough University, 2020
Role and employer	Research and Insights Manager, Science and Technical department, International Tennis Federation
What I do	The Science and Technical department are focused on protecting the nature of the game and encouraging innovation. We run several programmes to ensure quality and consistency across tennis balls, surfaces, and electronic line calling systems from our lab in London and on site at tennis tournaments. My role within the team is to research and develop new tests for tennis equipment and technologies. I use tennis-rich datasets such as ball and player tracking data to assess how the game is being played and address areas where more testing and regulation may be needed. To achieve this, I work closely with industry and academic partners to design, develop, and fabricate new test devices when necessary.
Past employers	Research Associate, Loughborough University
Academic interests	Sports equipment, sports surfaces, data analysis
Favourite sports	Football, tennis, rugby, cricket (all!)
Advice for students	Supplement your studies and take advantage of events such as sports engineering seminar days and conferences to help build your network.

(*Continued*)

Table 1.1 (Continued)

Dr Marcos David Moya Bencomo

Undergraduate degree	BSc Biomedical Engineering, Tecnologico de Monterrey, 2008
Postgraduate degree	MSc Sports Engineering, Sheffield Hallam University, 2010
Research degree	PhD Biomechanics, University of Manchester, 2018
Role and employer	Chief of the Sports and Health Innovation Engineering Laboratory and Design (SHIELD) and Professor at Tecnologico de Monterrey
What I do	In my role at SHIELD, I focus on creating technologies that help prevent and manage sports injuries. I am committed to making these technologies accessible to everyone, irrespective of their budget, to ensure that advanced healthcare solutions are available to all. My work in biomechanics is not only geared towards enhancing athletic performance in an ethical and fair manner but also aims to improve the quality of life for patients with locomotor pathologies.
Past employers	University of Baja California
Academic interests	Injury prevention and treatment, technology in sports medicine
Favourite sports	Cycling, swimming, and triathlon
Advice for students	Get actively involved with associations and organisations related to your field of interest. Take the time to listen and understand their needs and challenges. By doing so, you will not only develop valuable networking skills but also gain insights that can help you propose innovative ideas and solutions to address those needs. This proactive approach will enhance your learning, build your professional reputation, and open opportunities for collaboration and career growth.

The following metrics (from statistica.com) on the revenue of some global sporting goods markets highlight the scale of the industry:

- Sports equipment[5] was ~$123 billion in 2018, ~$172 billion in 2024, and is predicted to exceed ~$227 billion by 2029 [65].
- Sports apparel and swimwear was ~$64 billion in 2010, ~$142 billion in 2022, and is predicted to be ~$149 billion in 2023 [66].
- Athletic footwear[6] was ~$50 billion in 2018, ~$51 billion in 2022, and is predicted to be ~$65 billion by 2028 [67].
- Fitness/activity tracking wristwear was ~$4 billion in 2020, ~$6 billion in 2022, and is predicted to be ~$11 billion by 2028 [68].

With projected growth across many areas of sporting goods, sports engineers have many exciting opportunities.

1.5 Chapter summary

This chapter introduced the topic of Sports Engineering. The rest of this textbook will focus on equipping students with the technical knowledge underpinning Sports Engineering. This knowledge will leave students well equipped to become responsible Sports engineers who positively address the important topics and challenges facing them.

1.6 Suggested further reading

The following suggestions for further reading go deeper into the topics introduced in this chapter:

- *Advantage Play: Technologies that Changed Sporting History* [1]
- "Sports Engineering: an unfair advantage?" [14]
- "Sports Engineering | EBSCO Research Starters" [69]
- "Sustainable, inclusive, innovative: the role of engineering in sport" [15]
- "Engineering better sports venues: designing for athletes, spectators, and the environment" [16]

Notes

1 https://www.sportsengineering.org/
2 https://www.springer.com/journal/12283
3 https://journals.sagepub.com/home/pip
4 https://www.isss.ski/
5 In scope: Outdoor and indoor sports equipment, gym/exercise equipment, and protective gear and other accessories. Out of scope: Sports apparel and footwear.
6 In scope: Functional athletic footwear (e.g., for tennis, basketball, soccer, running, etc.). Out of scope: "Athleisure" footwear (everyday shoes with athletic design); work and safety shoes; ski boots.

2 Developing sporting goods

2.1 Introduction

Developing new sporting goods can be a lengthy and challenging process. A sports engineer would likely touch on almost all the content of this textbook when developing a new product, which can feel daunting. Thankfully, there are countless design theories and methodologies that can help structure this process, guiding us in the steps we should take, and how we should apply them through design [70]. Those familiar with design may have experienced "Design for X" (DfX), e.g., Design for *Assembly*, Design for *Manufacture*, or even Design for *Additive Manufacture* [71], "The Theory of Inventive Problem Solving" (TRIZ) [72], or the widely taught mechanical design processes from Ullman [73], or Ulrich, Eppinger, and Yang [74]. In this chapter we consider design as a structured problem-solving process, evaluating the steps required, and considering examples of how these steps might be applied to sporting goods.

We acknowledge that many of the best examples of sporting goods did not arise from the types of structured design process we see in Figure 2.1, and that experienced sports engineers might prefer to rely upon their intuition, gathered through years of discipline-specific practice. Nonetheless, every experienced sports engineer will be familiar with the structure and phases described within this chapter. This content might be considered the building blocks of successful sporting goods design. Here, we present the steps sports engineers should consider when searching for solutions for a defined problem.

2.2 Design as a process

Design Thinking is a multi-stage systematic design methodology that can be applied to various challenges. The five-stage process shown in Figure 2.1 originates from Stanford University. Each tile shows a phase filled with potential tasks that help drive the sports engineer towards a successful outcome. This is one of many design methodologies and is often shown in a simplified, sequential manner. In practice design processes tend to be more iterative. Regardless of the design methodology, our success is dependent upon our understanding of the problem. A sports engineer must focus attention on understanding all aspects of their design challenge to ensure that their proposals are fit for purpose.

DOI: 10.4324/9781003411086-2

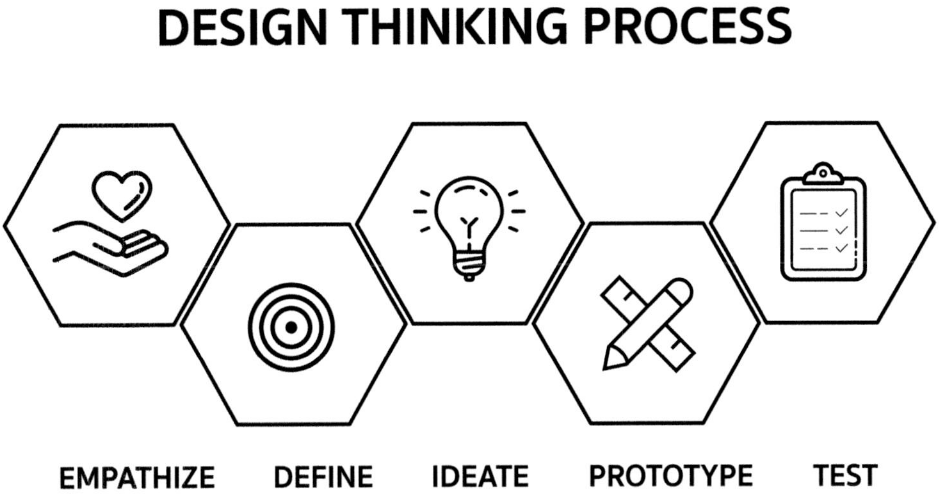

Figure 2.1 Illustration showing the five stages of the Design Thinking process: (1) Empathize, (2) Define, (3) Ideate, (4) Prototype, and (5) Test. This process outlines a structured approach to problem-solving.

Box 2.1

Behind each of the five tiles in the **Design Thinking** process in Figure 2.1 are tools and potential actions for the sports engineer to consider. The stages can be expanded as follows:

Empathize

Develop a deeper understanding of the challenge. Look to understand the user within the context of this challenge.

Define

Look to articulate the challenge to be solved. Work to identify the right problem and create an actionable problem statement.

Ideate

"Brainstorm" potential solutions and start to select and explore viable candidates. Produce many diverse solutions to solve the problem statement.

Prototype

Explore how well the developing solutions function. Build sample models that might represent any element of the design to explore function and form.

Test

Via testing, engage in a continuous short-cycle innovation process to repeatedly collect feedback from users to refine the problem, the solution, and their needs.

In this chapter we will see that the structured design process allows us to define the **requirements** of the product, and through these the overall **function** becomes clearer. This will be the primary purpose of the design output, and the answer to the question "what does it do?" This might seem like a trivial task, but the *real* function of a design and the perspective of its users is often misunderstood. There are countless examples of innovative products with sound engineering and seemingly great potential that fail to gain wider adoption.

Box 2.2

The **YikeBike** was a bold attempt to reinvent urban transport. It was a lightweight, folding electric bicycle inspired by the penny-farthing. Designed in New Zealand and launched in 2009, it gained global attention, won design awards, and promised a high-tech, portable alternative to bikes and scooters. It was listed as one of the Top 50 Inventions of 2009 by *Time* magazine [75]. A 2001 report from Manaaki Whenua (Landcare Research) highlights the perceived eco-potential of the YikeBike [76].

Early interest in the YikeBike was strong, but the product struggled to achieve mainstream success. Three key challenges held it back:

- **High price point:** Early models retailed for several thousand dollars, which was more than most electric bikes or scooters.
- **Limited practicality:** The riding posture and steering felt unfamiliar to many users, and its top speed and range were modest.
- **Niche appeal:** While it looked futuristic, the design did not suit most people's daily commuting needs.

As competitors offered simpler and cheaper options, interest declined. Since 2019, YikeBike have been seeking a merger and acquisition deal, with all technology, design, patents, intellectual property, tooling, stock, and manufacturing systems for sale.

Sports engineers must often balance the function of the product with its **form**. The latter depends upon material, manufacturing process, assembly and production, function, human interaction, aesthetics, regulation, certification, and more. The **means** is how we perform this balance. This is the bulk of the design challenge and brings opportunity for innovation. Thankfully, finding novel and improved ways to perform well-known functions is where sports engineers excel. In golf research, for example, sports engineers have determined player behaviours that contribute to reduced performance [77], helping to define an innovation space for novel products to explore (Figure 2.2).

2.3 The sporting goods design process

The overall aim for sporting goods design is often clearer than for other products. For example, sports engineers creating high-performance footwear for athletes know their performance objectives well. Unfortunately, this does not make sporting goods design easy. One of the main challenges to overcome is the specifics of how the athlete will use the product. High-performance equipment is often used at the outer reaches of human physical capacity. As a result, users are prone to error (unintended misuse) or to changes

(a) (b) (c)

Figure 2.2 Photos of golf putters, which are sporting goods with a specific and well-explored function. The wide array of product forms and means in which this function has been realised shows the incredible innovation of sports engineers.

in use behaviour, which can be hard to capture during the design phase. Endurance runners, for example, will change their gait as their muscles tire, which might provide different loading patterns to footwear than were otherwise expected [78].

In response to this specific need, a bespoke sports equipment design process model has been introduced [79]. A key feature of this model is user (athlete) input. The sports equipment design process model (Figure 2.3, [79]) highlights the cyclical nature of design and centres the user in every design phase. Despite the visual difference to other models (Figure 2.3), each of the core phases is represented. This model highlights the most important feature of sporting goods design – the user interface – and shows continual user involvement to embed this within the design process.

In this chapter we consider the key stages outlined in the methodology presented in Figure 2.3, as well as the **Stage Gates** required at the conclusion of each phase.

> Box 2.3
>
> The **Stage Gate** (or Phase Gate) is a point at which the design process pauses, and a decision is made on whether to progress to the next stage. Stage Gates are used to ensure that a project is *ready* to move on to the next stage, as well as to ensure buy-in from design teams and clients. In sporting goods design, this is a key point to engage with the user(s) and ensure that the developing solution meets their needs.

2.3.1 Research

The goal of the research phase is to gather and analyse all the information that helps us understand the problem, the target users, and the context in which the outcome will exist. This phase ensures the process is grounded in tangible evidence and is appropriate for our users. The target user group can differ in many ways. For example, the needs of recreational skiers will differ to elite racers. Research can encapsulate many actions, and the sports engineer must strike a balance of completing enough research to understand the challenge, but to not go so far that the project becomes unwieldy. In a commercial project, time and budget will likely be the key factors that determine the scope of the research phase. To prevent this phase from growing to become unmanageable, there are several techniques, processes, and defined outcomes that can help the sports engineer focus and structure their time effectively. Seven of the most common activities are highlighted in Figure 2.4.

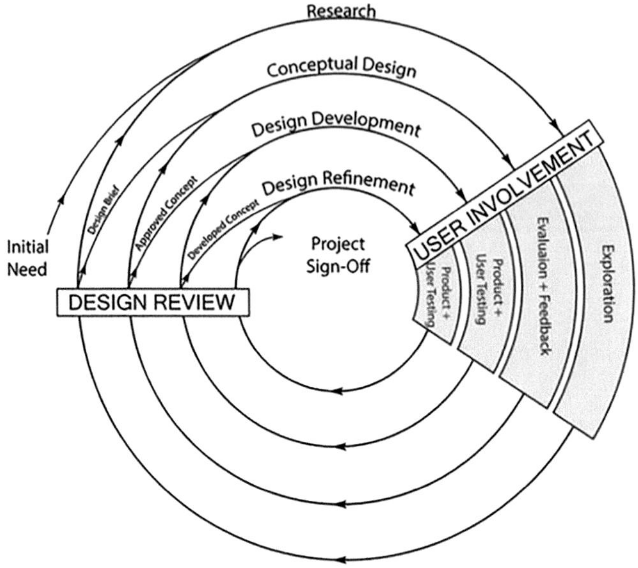

Figure 2.3 Schematic of the design process model for sports equipment design of Wilson et al. [79]. Licensed under CC BY 4.0.

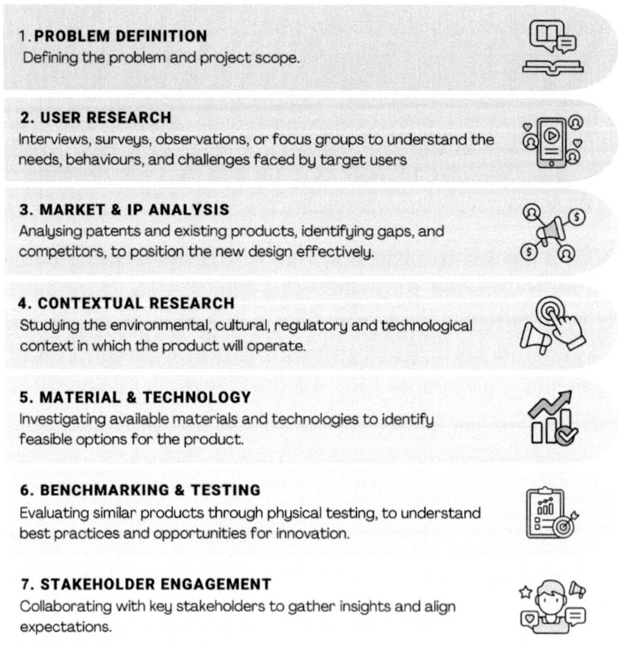

Figure 2.4 Illustration showing some key activities undertaken in preliminary research. The list is not exhaustive but provides a framework from which to develop a better understanding of the product and its design environment.

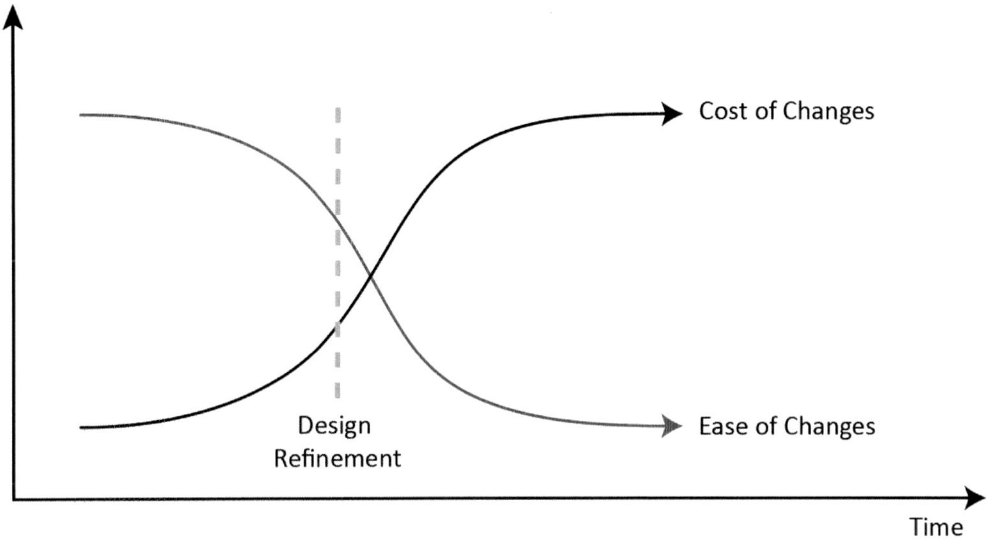

Figure 2.5 Illustration showing the effect of time on making changes within the design process. The time axis is arbitrary (and somewhat project-dependent). As an indicator, an approximate location for the design refinement phase is shown. After this point, as manufacturing readiness increases, the cost of changes grows rapidly.

This research phase defines the bulk of the project definition, and successful research will make the latter phases of the design project easier to manage. Experience shows that the cost of making design changes increases as the project progresses. Figure 2.5 shows the effect of time on the cost implications and the ability to make design changes. The design refinement phase is highlighted. The crossover point typically happens around deployment and production, after which changes are increasingly hard to make. For example, making changes to the shape of a bike helmet is much simpler during the conceptual design and prototyping phases than when moulds have been created for production. Even worse would be if the helmet has been released, and customers are reporting issues with comfort.

The findings from the research phase of a design process are typically summarised in reports, personas, journey maps, and **requirements specifications** to inform and guide the following design process phases.

Once the functional needs are understood, the sports engineer can explore how to meet them. This includes looking at materials, manufacturing processes, and other factors that influence the design. They will continue analysing existing products, exploring design context and regulations, and finding ways to make performance improvements. These improvements can come from changes in user needs (i.e., a market "pull") or from new technology in materials, manufacturing, or design (i.e., a technology "push").

Box 2.4

The **User Requirement Specification (URS)** is a list that details the needs of the product from the users' perspective. These *may* be split further into mandatory and *possible* elements hence providing some hierarchy to each element of the function. This is typically created early in the design process and can help create user or client buy-in of the project and its likely future direction.

To register (protect through patents, e.g., Figure 2.6 [80]) a new design, it must not only demonstrate an inventive step and be tangibly manufacturable, but it must also be sufficiently novel and cannot infringe upon existing protections. This means that time should be spent exploring existing design registrations. Through this, the sports engineer can identify what has gone before, what is protected, what is coming to the end of its protected period, and where opportunities for innovation might be found.

Box 2.5

The purpose of **design registration** is to protect intellectual property, the knowledge created through the design process. Registration forms a legal barrier, preventing the design from being copied and reproduced within the markets for which protection is obtained for a set period.

Ensuring that existing protections are identified is a key step in design, and there are many examples within sporting goods design where embarrassing (and often expensive) infringements have occurred. The risk is so high that many design companies will subcontract this research element to specialists to help shape their understanding of the intellectual property landscape. Once a sporting goods brand has produced a new design, they may also seek to protect it, so others cannot copy it.

In addition to intellectual property, sporting goods often need to meet explicit **standards** and **regulations** to be sold, or to be used within competition. In terms of standards, sporting goods design might be considered less *strict* than other contexts (e.g., electronic products or building design codes), yet most sports have regulations and requirements for equipment to be used within competition. Standards typically relate to safety, whereas regulations are often broader, covering aspects such as performance, officiating, and safety. The number of standards in sporting goods is continually reviewed and refined, and where there are none, sports engineers may choose to use those from neighbouring design spaces to help inform the design process. The topic of standards and regulations is covered throughout this textbook, including in Chapters 5 and 8.

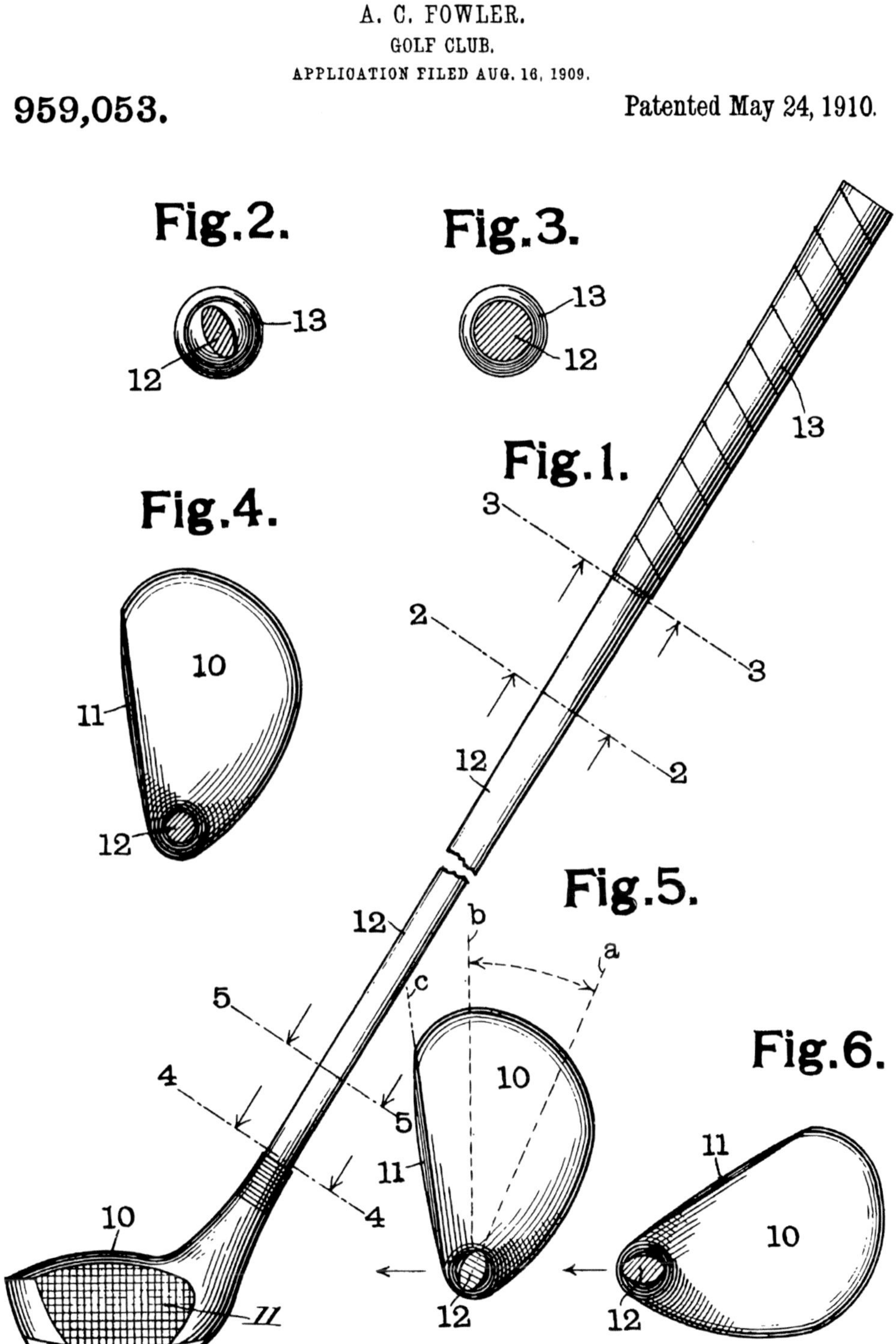

Figure 2.6 Illustrations of golf club designs from a vintage patent from 1910. The patent number is US959053A [80].

Box 2.6

Standards are documents that describe technical specifications to be used as rules, guidelines, or definitions, to ensure that materials, products, processes, and services are fit for purpose. These can be used during the design phase to describe mandatory requirements. Once a design is finalised, products are submitted for **certification** against these standards, proving that the design meets the specified requirements and permitting the product to be used or sold within a specific market or context. Standards can support international trade and increase customer and user confidence in a sports product.

Towards the end of the research phase, the sports engineer will typically create a **Product Design Specification (PDS)**. This document helps quantify all the requirements for the design. It provides a guideline to help create and evaluate concepts and defines the criteria by which a final design outcome will be judged. The PDS should be made visible to all stakeholders and will become an agreed (typically tabulated) statement of design intent. The PDS sits as a point of reference for all aspects of the design, including form, function, materials, usage, etc.

Ultimately, the purpose of the research phase is to convert a collection of broad, initial documents (the brief, market statements, intellectual property searches, the user requirements list, etc.) into a well-defined and measurable list of requirements. It is important that the PDS is comprehensive and clear and that all stakeholders agree with its contents. Poorly defined requirements inevitably lead to challenges later in the design and production phases (see Figure 2.5). As such a vital document, creation and agreement upon it often forms the **Stage Gate** for the project to pass through to reach the **conceptual design phase**.

Box 2.7

A **Product Design Specification** is a document that contains all the requirements, constraints, and specifications that a new product must adhere to. It should be a clear list with detailed information outlining every aspect of the design and have SMART criteria (Specific, Measurable, Achievable, Realistic, and Testable).

The PDS should be comprehensive. The following **PDS checklist** is provided as a starter/minimal set of considerations, but each project varies:

- Performance requirements
- User requirements
- Aesthetic requirements
- Materials
- Manufacturing and assembly
- Ergonomics (including form and mass)
- Stakeholder and market requirements
- Cost and budget

- Maintenance and support
- Distribution (scale, packaging, shipping)
- Standards and regulations
- Testing and validation
- Environmental conditions
- Lifecycle and environmental impact

2.3.2 *Conceptual design*

Following research, the project moves into a concept generation phase. Here, the **means** of *how* the requirements will be met is explored. This is the most creative and problem-solving of the design phases. It requires divergent and open thinking to define and explore a wide array of possible solutions.

Depending on the design challenge, this phase can range from fun and engaging to daunting and challenging. Componential Theory [81] (a branch of psychological research considering cognitive processes) describes that the main components of creative problem-solving are products of:

- Intrinsic task motivation
- Domain relevant skills
- Creativity-relevant processes
- The social environment

This can be reassuring, as each of these components may be studied and improved upon. In this section we assume that you are motivated, are developing appropriate skills in sports engineering, and are working in a supportive environment. We therefore focus upon creativity relevant processes. The process of creating ideas is known as **ideation** (as outlined in the "Ideate" tile of the Design Thinking process in Figure 2.1), and there are many techniques available. Ideas can even be inspired by research exploration, from nature [10, 21] or even generated using procedural artificial intelligence tools [82]. Sports engineers should spend time familiarising themselves with various techniques before focusing on methodologies that suit their users, preferred processes, or their workplace.

Sports engineers will typically conclude the conceptual design phase through **concept evaluation and selection**. This can be considered the **Stage Gate** for this phase. This process requires a mindset shift, from the divergent, broad-thinking exploration of ideas to a convergent, structured process for narrowing down and identifying the "best" option(s).

Selecting the "best" idea from viable candidates can be tricky, and sports engineers must ensure their biases and preferences do not influence this critical decision. To improve this, concepts are typically taken through a *screening* phase (to ensure all mandatory elements of the PDS are met) before being selected using a *scoring* method (to quantify how well these elements are met). The PDS (either in its entirety or selected elements thereof) can be used for both the *screening* and *scoring* processes. Whilst screening is typically binary, scoring can also be subject to bias, so it is good practice to ensure that project stakeholders are engaged in this

(a)

(b)

Figure 2.7 Brainstorming (a) is a group activity, typically centred around a white board or large notepad to capture the ideas generated by participants. Sketch-based ideation (b) is a process through which designers can explore design form and function and communicate developing ideas with stakeholders.

Box 2.8

The following overview lists techniques that can be implemented during **ideation**. Sports engineers might also choose to blend ideas together, or devise their own ideation techniques:

Divergent thinking techniques

- **Brainstorming** is a group activity designed to encourage free-flowing idea generation. Participants suggest as many ideas as possible without judgement, fostering creativity and collaboration (Figure 2.7a).
- **Mind mapping** is a visual tool to explore and organise ideas. Starting from a central concept, branches are drawn to represent related thoughts. This helps sports engineers see connections and expand their thinking.

Visual and physical techniques

- **Sketching** is a quick and low-cost way to put ideas on paper. This method allows sports engineers to explore and refine concepts visually without advanced tools (Figure 2.7b).
- **Prototyping** involves creating basic, tangible models of ideas. Prototypes can be physical or digital and are valuable for testing and iterating on concepts early in the process.
- **Morphological analysis** breaks problems into key components and systematically mixes and matches options to generate innovative solutions.

Collaborative ideation techniques

- **Co-creation** means working directly with others, such as users, peers, or stakeholders, to develop ideas. This ensures the concepts are relevant and grounded in real-world needs.
- **Storyboarding** is a visual storytelling method that maps out a sequence of events or user interactions. It helps sports engineers visualise ideas in action and identify potential improvements.
- **Affinity mapping** helps to organise and cluster ideas by theme. This technique is helpful for identifying patterns and prioritising the most promising concepts.

Playful ideation techniques

- **Constraints-based ideation** uses specific limitations (e.g., materials, budget, or time) to push creative thinking. The technique helps sports engineers to innovate within real-world constraints.
- **Serious play** is a hands-on activity where sports engineers use physical building blocks to rapidly create and communicate ideas. This method fosters creativity, teamwork, and problem-solving in a fun way.

Table 2.1 An example concept selection matrix for the selection of the primary layout/function of an automatic cricket ball launcher (numbers are arbitrary to show process)

		Concept 1		*Concept 2*		*Concept 3*	
		Dual-wheel		*Single-wheel*		*Compressed gas*	
		"control master"		*"spin centric"*		*"high velocity"*	
Selection criteria	*Weight (%)*	*Rating (1–10)*	*Weighted score*	*Rating (1–10)*	*Weighted score*	*Rating (1–10)*	*Weighted score*
Market differentiation	10	6	0.6	6	0.6	9	0.9
Ease of use	15	8	1.2	9	1.35	6	0.9
Launch conditions	20	9	1.8	8	1.6	7	1.4
Launch consistency	10	6	0.6	7	0.7	9	0.9
Materials and durability	10	8	0.8	8	0.8	9	0.9
Safety	10	8	0.8	9	0.9	6	0.6
Production feasibility	10	7	0.7	8	0.8	6	0.6
Sustainability	10	7	0.7	7	0.7	6	0.6
Aesthetic and ergonomics	5	9	0.45	8	0.4	7	0.35
Total score	100		**7.65**		**7.85**		**7.15**

phase. Table 2.1 shows an example selection matrix being used to select concepts. As can be seen, various attributes from the PDS are used, and each are assigned a weighting to define their relative importance. The "best" solution is the one with the highest weighted score.

One of the benefits of using a structured system is that concepts are ranked as well as selected, which helps to identify further options should the selected concept become unworkable for any reason. The process also highlights which concept scored best in each category. With this information a new concept combining all of these "best" attributes can also be considered.

Box 2.9

In the **concept selection phase**, the design team narrows down various potential ideas into a chosen solution. This step is crucial because selecting the right concept is key to the project's success. However, sports engineers can sometimes fall victim to conscious or unconscious biases, which makes it essential to use structured and objective methods for concept selection. Choosing the "wrong" concept can lead to costly mistakes, wasting time and resources.

Equally important is ensuring that the chosen solution resonates with the user group. Even a technically valid concept is worthless if it fails to meet user needs or expectations. Engaging users early and obtaining their support during this process is key to avoiding missteps.

Through ideation techniques and structured concept selection, sports engineers can create and *justify* a proposed solution to carry forward to the design development stage. The solution they create here may need protection to prevent others from capitalising on the innovations being developed. In the sports context, it is particularly important that user agreement has been sought, and the concept should be analysed alongside these stakeholders to determine any potential weakness and areas where further research is required. Through this, a structured process of design development can begin.

Box 2.10

The timing of a patent application is critical. If we apply too early, the idea may be too underdeveloped to protect effectively. Unfortunately, applying too late often risks losing the opportunity altogether.

Patent-related activity should generally begin during the research and conceptual design phases, with a focus on investigating novelty through prior art searches and analysis of existing intellectual property. At this stage, teams can evaluate whether their idea is genuinely new and potentially patentable.

The actual filing of a patent (particularly a provisional patent application) is typically done once the design has progressed further. This often occurs during the design development or refinement phase, when the innovation is defined clearly enough to be described in detail. A provisional application secures a priority date and provides 12 months to further develop the idea before committing to a full (non-provisional) filing.

Most importantly, a patent must be filed **before any public disclosure** of the invention. This includes publishing, presenting, exhibiting, or selling the product. In many jurisdictions (such as the European Union (EU)), public disclosure without a prior filing can disqualify an idea from being patented. This makes it essential to consider patent strategy early, especially when discussing concepts with potential users, clients, or focus groups.

2.3.3 Technology readiness

When designing new products, and selecting supporting technologies, it is important to know not just whether something could work, but how much confidence the design team has in the outcome. Technology Readiness Levels (TRL) (Figure 2.8) use a nine-point scale that describes how mature a technology is, from initial idea through to full commercial application. The framework was originally developed by NASA but is now widely used across industry and academia to help understand development risk and to track progress. For sports engineers the TRL scale offers a useful logic check to see how close a solution is to being viable, and what kind of work needs to be completed before it can be launched.

The TRL scale can help to keep us grounded as we identify or develop the "next best thing". Each level requires stronger evidence of performance; hence the scale reminds us that even innovations that show incredible performance as a prototype in a laboratory environment (TRL 4) remain far from full commercial application. Academic research undertaken within universities and published in scientific papers is typically at the lower TRL (e.g., ≤ TRL 4). This means considerable effort and investments are often needed to progress an idea from a scientific paper through the TRL scale to full commercial application at TRL 9. In a similar manner, Manufacturing Readiness Levels (MRLs) use a nine-point scale that describes how mature a manufacturing process is. Sports engineers must consider where a technology or process sits on the TLR and MRL scales when

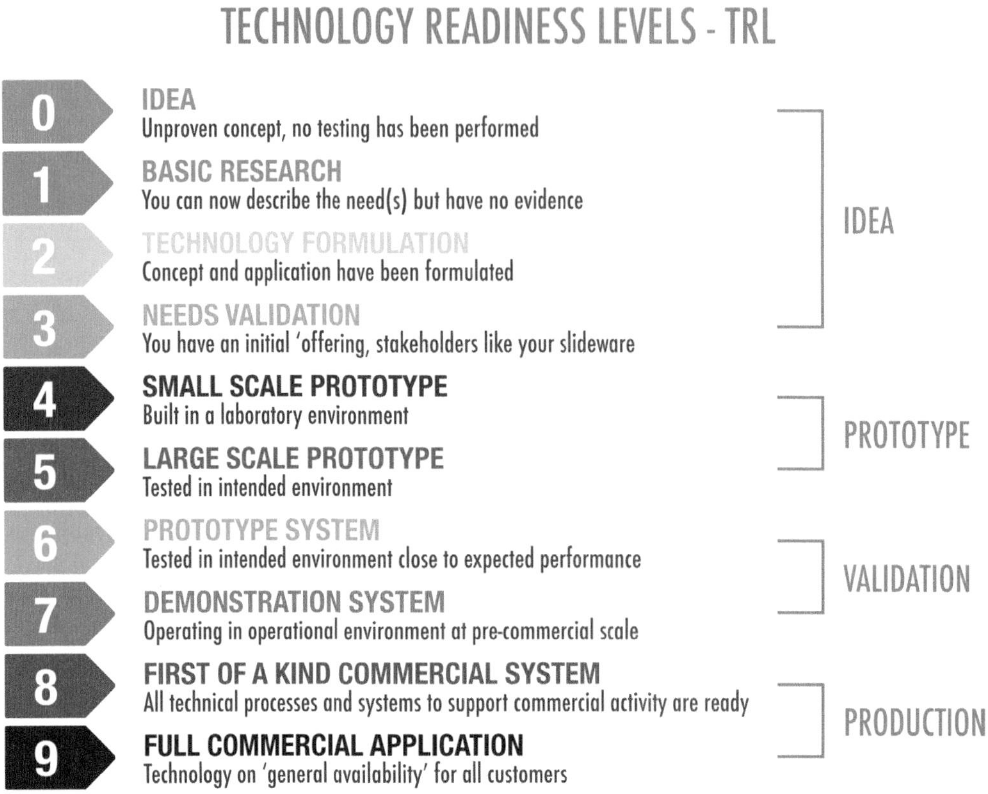

Figure 2.8 Illustration showing the Technology Readiness Levels, which are a nine-point scale that highlight technology maturity, from initial idea to full commercial application.

considering it for their design. Those at the lower levels are higher risk and will require more resource to progress to application. They may also bring higher reward as a more innovative sports product.

2.3.4 *Design development*

The design development phase is where sports engineers start to determine specific details for the product. In most cases, a concept has been selected for development, and the bulk of the work is in converting this skeletal, conceptual design into a prototype. Sports engineers need to focus on resolving technical challenges, optimising functionality, and enhancing the user experience while ensuring the solution aligns with project goals and constraints, such as budget, materials, and manufacturing feasibility. This means that active collaboration is critical in this phase, and the sports engineer must be both a leader and a team player. Prototyping and testing play a central role, helping evaluate concept performance, identify potential issues, and re-iterate quickly. The goal of design development is to create a solution that balances aesthetics, usability, and technical requirements, ensuring it is both desirable and viable for production. In many sports contexts, sponsored or ex-professional athletes (i.e., brand ambassadors) may be supplied with prototype equipment to test and provide feedback on.

This stage of the design will likely see an increase in the number (and complexity) of prototypes produced, whether as whole-concept visual and physical mock-ups or subsets of components, features, etc. that can be tested more tangibly. These prototypes are created to assess the validity of the developing solution. They may start as rapid, low-fidelity solutions through card or foam modelling to assess shape, form, and ergonomics. Later, digital models will be produced, presenting opportunities for virtual assessments of structure and performance through simulation. Digital models also permit rapid prototyping through additive manufacture (i.e., 3D printing), where near-final shapes can be created to assess form and function (Figure 2.9).

Figure 2.9 Digital models allow designers to produce high-quality renders and to rapidly prototype their developing designs through additive manufacture.

Box 2.11

The **Bill of Materials** is an agreed document that communicates the items (parts, components) that are required to create a product, as well as the material, finishing, and assembly requirements. This document must accurately reflect all elements of the design to ensure that costing is accurate, and that manufacture and assembly are efficient.

The developing solution will go through rounds of product testing to ensure that the desired functional and regulatory requirements can be met. Chapter 8 considers the testing of sporting goods in more detail. As the final product form develops, user groups should be re-engaged to interact with the product and provide further feedback.

At the same time, proposed manufacturing partners will be engaged to provide feedback on the design solution as it is finalised. A Bill of Materials will be created, alongside production volume and timescales. From this, preliminary costing can be confirmed, and the value proposition of the solution be determined.

The **Stage Gate** concluding the design development and moving to the design refinement is typically referred to as a **Design Review** or **Critical Design Review** (**CDR**). This checkpoint allows the sports engineer to assess whether the concept is ready to transition from development into the refinement phase.

Box 2.12

During the **Design Review**, the design team evaluates key aspects such as functionality, feasibility, and user feedback. They check if the design meets all core functional requirements, if the materials and manufacturing processes are technically feasible, cost-effective, and sustainable, and whether testing has validated the design proposal. Potential risks or unresolved issues are also identified and addressed. If the design successfully passes this review, it moves into the **design refinement** phase, where the focus shifts to fine-tuning and preparing the product for production. However, if issues arise during the review, the design may need to return to the **design development** phase for further iteration.

2.3.5 Design refinement and project sign-off

In the design refinement phase, the focus shifts to fine-tuning and finalising the design before it moves into production. This stage involves making the final adjustments to ensure the product meets all user needs, performance standards, and quality requirements. Sports engineers will assess and improve details like ergonomics, aesthetics, material selection, and manufacturability, ensuring the design is both practical and appealing.

Box 2.13

Design refinement and **design development** seem similar but have key differences. In **design development**, the focus is on transforming the chosen concept into a fully functional and feasible solution. This phase involves addressing technical challenges, iterating on prototypes, and refining the design to meet user and performance requirements. In contrast, **design refinement** shifts towards fine-tuning and perfecting the design, focusing on minor adjustments, optimising details for production, and ensuring a desirable user experience.

Design development solves large-scale challenges and evaluates core concepts, where refinement removes imperfections and ensures the design is production ready. In the development stage, design iterations may involve substantial change to a design, whereas in the refinement stage the focus is on making subtle adjustments to align the design with the project goals.

Within this phase, feedback on prototypes from users and production teams are crucial to resolve any issues and finalise the design for real-world use. The goal is to remove any potential flaws, streamline the production process, and ensure that the product is ready for production. The design should now be fully aligned with the initial vision, while being feasible, efficient, and well received by users. For sporting goods, this phase will see a continuation of user testing, wherein brand ambassadors take late-stage prototypes into the field (and even competition) to provide final feedback to the design team.

For complex assemblies and larger-volume manufacture, the refinement stage allows sports engineers to determine critical detail on production costs and provides a final chance for cost optimisation. Sports engineers will work with production partners through early manufacturing to streamline processes and material usage to minimise costs and maximise productivity. The final Stage Gate of the entire process is that of **project sign-off**, which concludes the design phase.

Box 2.14

The final **project sign-off** is likely to be a scheduled event, and thus the project is timebound by this date. Much like the Stage Gates between each phase, sign-off represents a specific point at which the design is approved. However, sign-off is the *final* approval stage, where the design is formally deemed complete and ready for production.

It involves a thorough review to ensure that all project objectives have been met, including functionality, design specifications, user needs, and budget constraints. At this point, all stakeholders, including sports engineers, product managers, and clients, verify that the final design aligns with the initial vision and meets quality standards. Any outstanding issues or required changes should be addressed before sign-off. Once the design is officially approved, it moves into the production phase, and the design project is closed.

If a systematic and Stage-Gated process has been used, the final sign-off should be straightforward. For less structured projects, or for those that have had limited stakeholder buy in, reaching final sign-off may be challenging.

2.3.6 Production and quality assurance

After sign-off, production and quality assurance play a critical role in translating the design into a reliable, high-performance item of sporting goods. The production phase encompasses the practical realisation of a design, beginning with considerations around manufacturability. Sports engineers must ensure that products can be efficiently and consistently mass-produced. This often means adapting prototypes to meet the constraints of tools, materials, and assembly processes. This is particularly important for sports technology, where products often combine mechanical components with electronics, and must endure repeated use in challenging environments. The transition from small-batch prototyping to mass production involves careful validation of processes, supply chains, and final assemblies to ensure consistency and repeatability. Manufacturing is covered further in Chapter 7.

Quality assurance plays a crucial role in ensuring the integrity, safety, and performance of a design. It typically involves thorough mechanical and environmental testing to confirm that the product works reliably in real-world conditions. This could be anything from resisting ultraviolet exposure and moisture to maintaining sensor accuracy over time but should have been highlighted through research and outlined in the PDS. In sports engineering, quality assurance also means meeting relevant safety and performance standards and regulations and includes user testing to assess how well the product performs in practice. For designs that include embedded software or app-based features, this process extends to validating the software itself. Ultimately, production and quality assurance go hand in hand, helping ensure that the final product is both manufacturable and dependable for the end user.

In 2023 the Sports Tech Quality Framework was introduced [83]. This aimed to address variability in sporting goods development, in particular digital technologies for sensing and instrumentation. It focusses on five main "pillars" of quality, as shown in Figure 2.10.

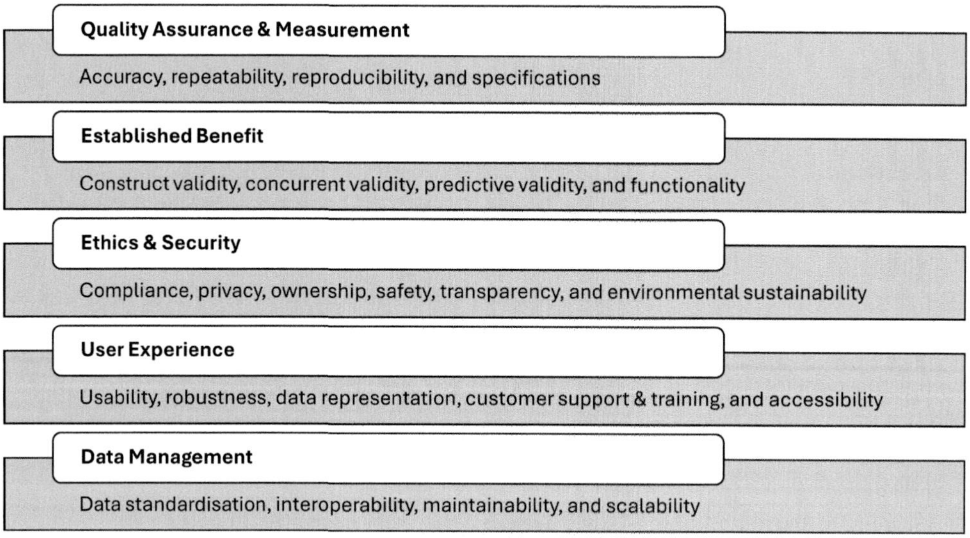

Figure 2.10 Diagram showing the five main "pillars" of quality from the Sports Tech Quality Framework [83].

The pillars present a comprehensive overview of topics that assist users, governing bodies, sports leagues, education providers, and start-ups in matters pertaining to development, launch, and procurement. A follow-up white paper [84] "Bringing the Framework to Life" introduced 23 infographics (Figure 2.11) to add further clarity and assist sports engineers in implementing the strategies proposed.

2.4 Case study: Sports wearables

Sports wearables are devices or apparel equipped with sensors and connectivity features to track and analyse physical activity and performance. They often sync with smartphone apps for detailed insights into sports activity (see Figure 2.12). Wearables offer benefits such as real-time feedback, enhanced performance monitoring, and long-term health tracking, making them valuable for both casual users and athletes. Their popularity has surged recently, driven by advances in sensor technology, the growing interest in fitness and wellness, and the integration of wearables with smartphones, health apps, and social media. They have become staples in activities ranging from running and cycling to team sports and strength training.

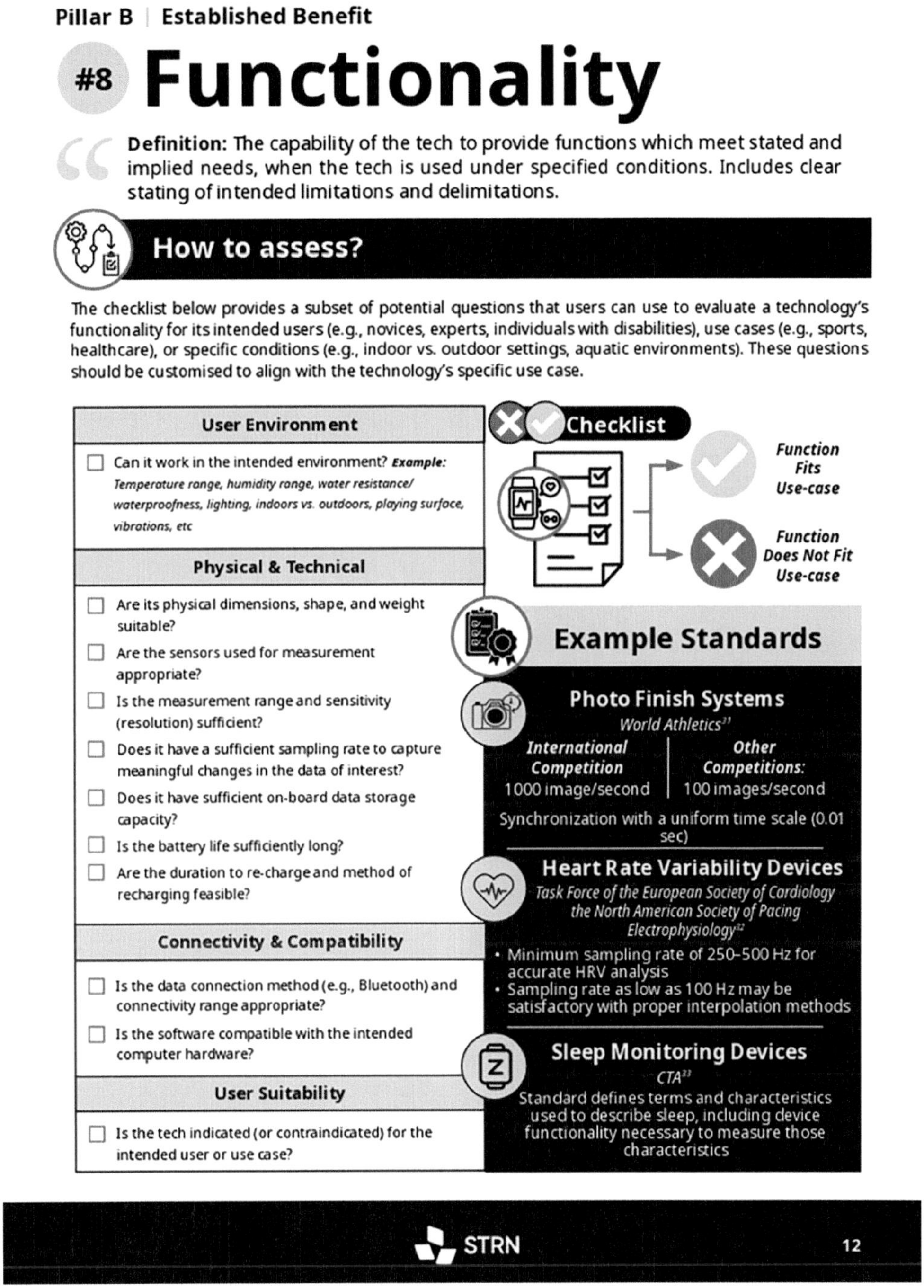

Figure 2.11 This infographic is a sample of the 23 introduced by the Sports Tech Quality Framework, highlighting sample questions that might be considered in assessing product functionality. Image reproduced from the Bringing the Framework to Life white paper [84] with the kind permission of the Sports Technology Research Network.

Wearables appeal to a broad audience seeking data-driven ways to improve their performance and wellbeing and for sharing progress with friends and family. Despite their popularity, these devices are often generic, covering various sports or activities with one set of sensors. In the case study that follows, an overview of the co-design of a wearable device for a specific activity (**sport climbing**) is considered. The case study follows the work of Mencarini et al. [85] and focuses on the design process and the rationale for the steps taken.

Despite the growth in interest (and sales) of wearable technologies, users often report challenges. These include discomfort during use, complex user interfaces, limited

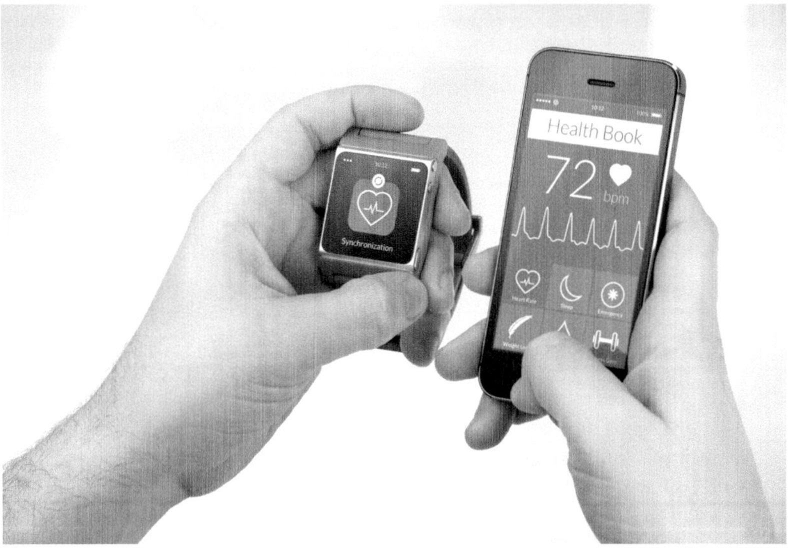

(a)

(b)

Figure 2.12 Sports wearables have grown in popularity and offer a wide array of features that can be synced with smartphone apps for real-time activity and health monitoring.

battery life, and difficulties with syncing data or general data compatibility. Each challenge presents a technological opportunity to be exploited through design, yet this project explored the challenges further rather than immediately moving to remediation. In this challenge exploration, the following queries were posed:

- Would climbers want to use wearables?
- If so, for what purpose?
- What design attributes are required for a product to be accepted by climbers?

Box 2.15

Sport climbing involves climbing pre-equipped routes on rock faces or artificial walls (Figure 2.13). The aim is to reach the top of the climb, often within a set time limit, using technique and strength.

Unlike "traditional" climbing, where climbers place their own protection (such as cams or nuts), sport climbing relies on fixed protection to reduce the risk of falling. It includes various climbing disciplines, such as lead climbing, bouldering, and competition climbing, with the latter now an Olympic sport.

The design process was biased towards understanding the value proposition of any new device through a comprehensive research phase.

Understanding the importance of user engagement within sporting goods design, the project employed a multiphase co-design process to develop a wearable device for coaching novice climbers. The process was broken down into five key steps, as shown in Table 2.2 [85].

Through research, they found that climbers were initially hesitant to use wearable technology because of perceptual conflicts around how and why they enjoy climbing. By conducting focus groups and engaging with climbers' perspectives, three key values were identified as essential design principles for developing more acceptable wearables (Figure 2.14).

The project identified that beginner climbers should be the primary user group for this type of device, due to their openness to support and their developmental needs. A proposal for a device that enhanced communication between instructors and novices was created. The device allows instructors to provide guidance on posture and movement via haptic feedback, minimising disruption to the climber's focus and overall experience. The proposal was designed to respect a need for unobtrusive, supportive tools that preserve the immersive flow of climbing.

This case study provides a snapshot of the challenges involved in developing sporting goods. Across the five design steps, it becomes clear how much effort is required in early-stage development, particularly in understanding user needs and forming a strong concept design. Early-stage design plays a critical role in setting the direction for innovation, and this example shows just how much groundwork is needed to do so successfully. Moving this project from prototype (TRL 4) to a market-ready product (TRL 9) would involve refining the design using broader user feedback (TRL 5). Testing could expand to include climbers in varied settings, especially outdoors, to ensure the device performs reliably in different conditions (TRL 6). This phase would also help gauge interest and clarify how the product might be adopted.

Design refinement would likely need to focus on reliability and ease of use. Clear communication and seamless connectivity (especially in both indoor and outdoor environments) would be essential. Longer-term evaluation could explore device efficacy, e.g., how it influences climbers' experiences (such as its impact on skill development or confidence) over time (TRL 7). Before a full launch, a small pilot release might be considered to provide valuable real-world feedback (TRL 8). This final testing phase would reveal

(a)

(b)

Figure 2.13 Two photos showing examples of sport climbing.

Table 2.2 Five steps used by Mencarini et al. [85] to design a sports wearable for sport climbing

Step	*Phase*	*Process*	*Key findings*
Understanding climbers' attitudes towards wearable devices	Research phase	Focus groups with climbers and designers to identify the factors that influence climbers' acceptance of wearables.	The culture and context of climbing leads to a general reluctance to technology. 6/15 climbers were explicitly against technology use. Overall, technology should: • Support, not substitute • Be handy but not obtrusive • Maintain a social image of competence • Be minimal in scale and presence • Be reliable and useful
Opportunities for wearables in climbing	Research phase/ conceptual design phase	Design workshop with design professionals practising outdoor sports creating outline concepts.	Most concepts focused on communication between climbing partners: • Sonification – using music to trigger changes in technique • Emotion-matched T-shirts to convey heart rate, etc. between climber and belay partner • Post-climb feedback through movement measurements • Attention monitoring (of belayer) to improve trust
Exploring specific technologies	Conceptual design phase	"Bodystorming" design workshop with climbers and designers in climbing gym, using physical movements to explore haptic stimulation.	Key findings through physical exploration were: • A need to provide well-located feedback mechanisms • Belayer has limited free movement – suggesting use of pedalboard • Use of multi-part necklace for directional and auditory messaging • Automatic status (traffic light) communication with haptic stimulus for change in status
Designing and implementing the technology	Conceptual design phase	Authors' design of a vibrotactile feedback prototype.	Authors created a functional prototype incorporating: • Multi-location haptic feedback elements to direct limbs and movement • Control and communication software (tablet-based) for belayer to operate and advise climber
Evaluating the design proposal (prototype)	Conceptual design phase	Feedback session with ten climbers and one guide using prototype to train within climbing gym environment.	Feedback showed positives in: • General acceptance of this as a useful training tool • Improvements over default training tools (such as laser pointers) Ambiguity in: • Where haptic devices should be placed • How best to send/receive signals Challenges in: • Interpreting the meaning of signals • Timing of signals • Control of signal sending

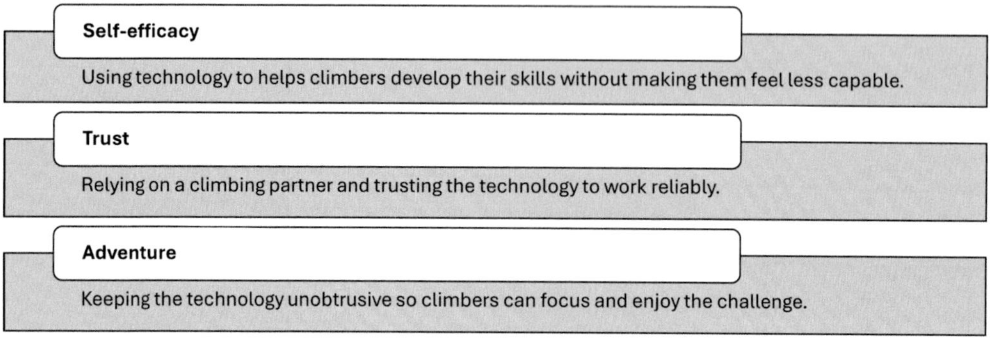

Figure 2.14 Diagram showing three key values considered to be essential design principles for developing more acceptable wearables for climbers, which are based on those identified by Mencarini et al. [85].

any remaining issues and guide final refinements. With these steps complete, the wearable would be well positioned to develop from a promising prototype into a practical, well-aligned tool for the climbing community (TRL 9). Such a process could be applied to develop sports wearables specifically for other types of climbing or activities like mountain biking, roller sports, or snow sports.

2.5 Chapter summary

This chapter has provided a broad overview of developing sporting goods. It began with an overview of basic design processes and their application. The importance of following a systematic design process was emphasised and each step in this process was explored. The chapter concluded with a case study on sports wearables for sport climbing. The basic understanding gained from this chapter should be sufficient for a budding sports engineer to see how development processes for sporting goods relate to the complementary topics covered in this introductory textbook. However, there is still more to be learnt about sporting goods development beyond the basics covered here.

2.6 Suggested further reading

The following suggestions for further reading go deeper into the topics introduced in this chapter:

- "Co-designing wearable devices for sports: the case study of sport climbing" [85]
- "Development and presentation of the first design process model for sports equipment design" [79]
- *Routledge Handbook of Sports Technology and Engineering*, 1st and 2nd editions [10, 86]
- *The Mechanical Design Process* [70]
- "A New Sports Tech Quality Framework White Paper: Bringing the Framework to Life – Sports Tech Research Network" [84]

3 Sports mechanics

An understanding of the basic mechanics of sport is key to sporting goods design and sports engineering in general. By understanding and applying mechanics principles sports engineers can develop tools to help athletes perform better. A good understanding of mechanics also allows engineers to make sport safer. For example, they can effectively design sporting goods, including protective devices, and better manage venues, such as snow sports resorts and bike and skate parks. Tests applied to further our understanding of sports mechanics were traditionally constrained to indoor laboratories. The emergence of smaller, portable camera systems and wearable devices has allowed for testing outdoors and during actual sport [18].

This chapter covers the fundamentals of sports mechanics. Basic mechanics topics are introduced and discussed in the context of sport and sporting goods.

3.1 Units

When working on mechanics problems, it is important to ensure consistency in units of measurement. This book uses the International System of Units, which are commonly known as SI units or the metric system. There are seven base units (see Figure 3.1). The first three base units (s, m, kg) are most relevant to sports mechanics, and the topics covered here. From these base units, those for other measurements can be derived. For example, velocity is obtained from the ratio of distance (m) and time (s). As such, the units for velocity are metres per second, or m/s. If a runner completed a 100 m race in 10 s, their average velocity would have been 10 m/s.

A prefix can also be used to indicate multiples or fractions of SI units, which helps with efficient and concise communication. Table 3.1 shows prefixes commonly used in sports mechanics. For example, a 10,000 m run is often referred to as a 10 km run. Further information on units can be found in ref. [87].

3.2 Newton's laws of motion

We start with Newton's laws of motion, of which there are three (Figure 3.2).

First law: The law of inertia.

If a body is at rest or moving at a constant speed in a straight line, it will remain at rest or keep moving in a straight line at constant speed unless it is acted upon by a force.

DOI: 10.4324/9781003411086-3

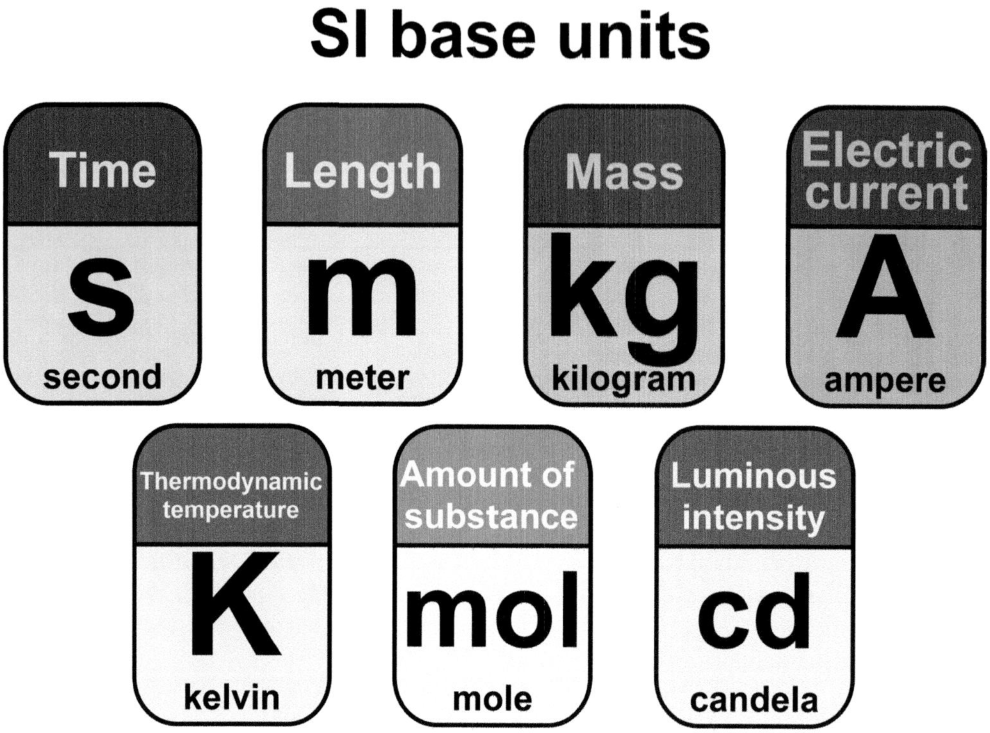

Figure 3.1 Illustration showing the seven SI base units.

Table 3.1 Table showing unit prefixes commonly used in sports mechanics

Prefix	*Symbol*	*Factor*	*Example*
giga	G	10^9	giga pascal* or GPa
mega	M	10^6	mega pascal or MPa
kilo	k	10^3	kilometre or km
centi	c	10^{-2}	centimetre or cm
milli	m	10^{-3}	millimetre or mm
micro	µ	10^{-6}	micrometre or µm

* Pascal is the unit of pressure, which is the ratio of force to area.

For example, a golf ball on a tee is at rest. If the golf ball is struck by a club, the resulting impact force will cause it to accelerate and hence no longer be at rest. Another example is a football resting on a penalty spot, which if kicked will accelerate towards the goal (see Figure 3.3a). These balls would not actually travel in a straight line at constant speed, due to the forces of their weight and air resistance acting on them (more to follow on this topic). The weight of a body is the product of its mass and gravity (g, ~9.81 m/s^2 on earth). In another example, imagine a skydiver jumping from a plane (Figure 3.3b). Following a period of gravity-induced acceleration, they would reach their terminal velocity. This terminal velocity is when the opposing forces of their weight and

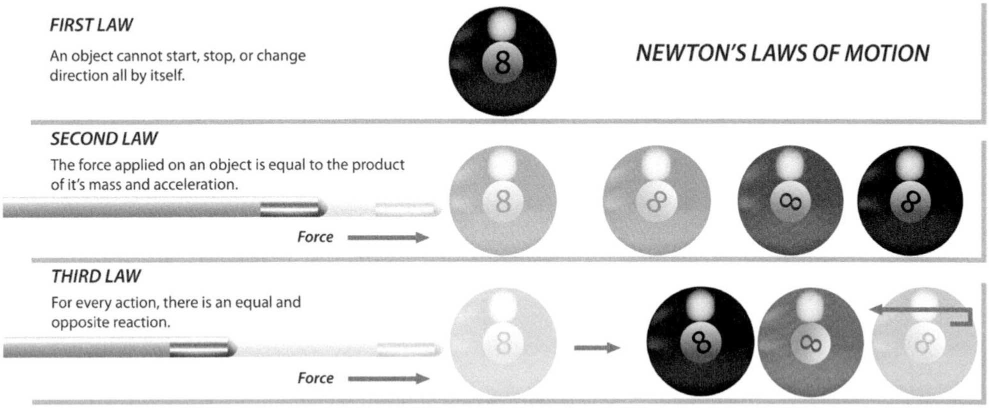

Figure 3.2 Illustration showing Newton's laws of motion.

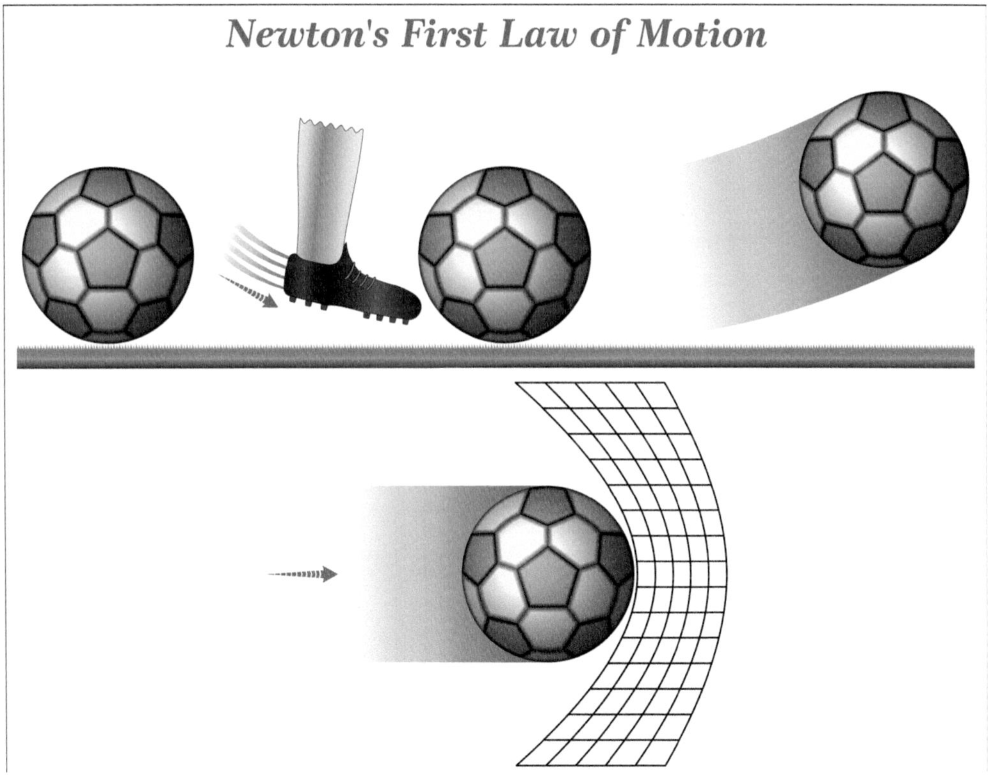

(a)

Figure 3.3 Newton's first law of motion illustrated by (a) a football being kicked and (b) terminal velocity during a skydive, with the weight and drag forces acting on the skydiver labelled.

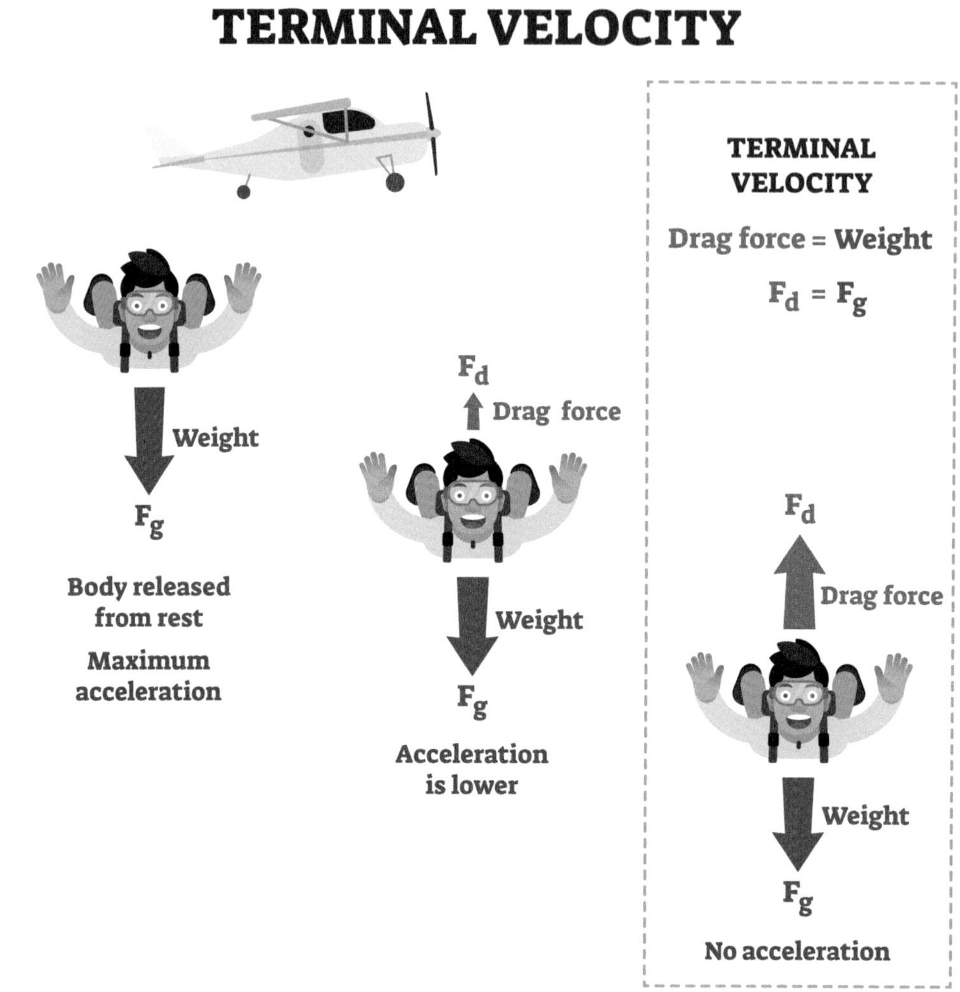

(b)

Figure 3.3 (Continued)

air resistance, or drag, become equal, meaning they will travel at a constant speed in a straight line (i.e., down towards earth).

Second law: Force and change in motion.
Force (F) is the product of mass (m) and acceleration (a).

or

$$F = ma \qquad \text{Equation 3.1}$$

and

$$a = \frac{u_f - u_i}{t_f - t_i} \qquad \text{Equation 3.2}$$

where u is velocity, t is time, and the subscripts f and i are final and initial, respectively. This means Equation 3.2 can be simplified to:

$$a = \frac{\Delta u}{\Delta t} \qquad \text{Equation 3.3}$$

where the Greek symbol Δ (delta) is used to indicate a change, which in this case is a change in both u and t. Indeed, acceleration is the ratio of change in velocity to change in time.

Let's return to our example of a ball being struck. When struck, the ball would change its velocity (from stationary). The size of this change depends on the ball's mass and the amount of force applied to it by the club or foot. This is why the mass of sports balls is regulated by the respective sporting governing body (more to follow on this topic), ensuring they behave as expected during play. Hence it is the amount of applied force that dictates the acceleration of a given type of sports ball.

> **Third law:** The law of action and reaction.
> When two bodies interact, they apply forces to one another that are equal in magnitude and opposite in direction.

This means the tee or pitch will apply forces (upwards) on the resting golf ball or football equal and opposite to their weight. When struck, they will apply forces on the club or foot equal and opposite to the striking force.

3.3 Forces, moments, and static equilibrium

3.3.1 Forces

A classic example of forces in sports engineering is those acting between the foot and ground during walking or running (Figure 3.4a). Such force is often termed ground reaction force (GRF). Figure 3.4b illustrates a typical vertical ground reaction force plot, as can be obtained from a person running over a force platform. Temporal means relating to time, and in this example the time has been normalised to the stance phase. Foot–ground contact starts at 0% stance and ends at 100%. The shape of the temporal force plot depends on various factors such as the mass and speed of the runner, their running style, and their footwear [88–90]. Indeed, many sports engineers work in footwear development. Key parameters can be obtained from such a temporal force plot, such as the loading rate, the impact peak force, and the active peak force. The mean loading rate can be obtained by fitting a straight line to the initial section of the plot from ground contact to peak impact force (i.e., ~0 to 20% stance). The active peak force is typically around two to three times body weight, depending on running speed.

3.3.2 Moments

A moment (M) results from a force being offset from a fixed point or axis:

$$M = Fd \qquad \text{Equation 3.4}$$

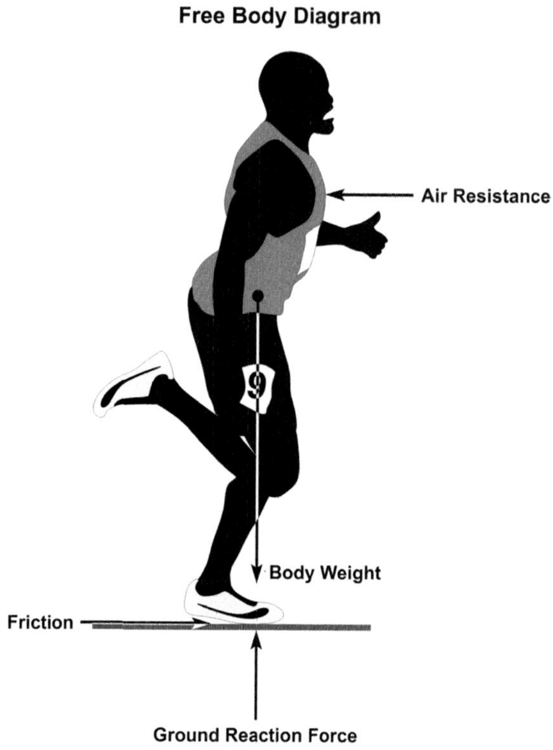

(a)

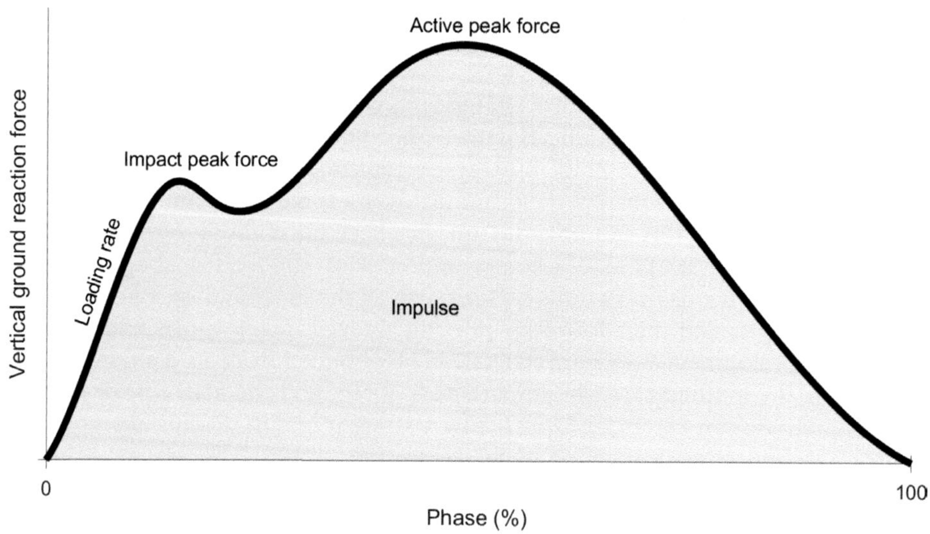

(b)

Figure 3.4 Illustrations showing (a) a free body diagram demonstrating the forces acting on a man running in a straight line, including air resistance, body weight, friction, and ground reaction force, and (b) a typical ground reaction force plot, with annotations showing the loading rate, impact peak force, active peak force, and impulse.

a) b) c)

Figure 3.5 Photos showing examples of moments in sport, (a) a diver on a diving board, their weight creates a moment about the fulcrum which causes the board to deflect, (b) a skateboarder "dropping in" on a ramp, as they shift their weight forward the board will pivot about its tail, and (c) a cyclist pulling on a brake lever, which creates a moment about its attachment point.

where d is the offset distance. Such a moment can create a turning or twisting effect on the body. There are many examples of moments in sport. For example, a diver standing on a diving board creates a moment acting about its fulcrum, causing it to bend downwards (Figure 3.5a). If their mass was 100 kg and they were a meter from the fulcrum, according to Equation 3.4 the resulting moment would be almost 1 kNm:

$$100\ kg \times 9.81 \frac{m}{s^2} \times 1\ m = 981\ Nm$$

Other examples include a skateboarder "dropping in" on a ramp by loading the front of the board (Figure 3.5b) and a cyclist pulling on a brake lever (Figure 3.5c).

3.3.3 *Static equilibrium*

Static equilibrium is when a body is at rest because the forces acting on it are balanced. In a two-dimensional (2D) problem (i.e., a scenario that is assumed to be 2D), this means the sum (Σ) of any forces, in both the horizontal (x) and vertical (y) direction, and any moments all equate to zero:

$$\Sigma F_x = 0 \qquad \text{Equation 3.5}$$

$$\Sigma F_y = 0 \qquad \text{Equation 3.6}$$

$$\Sigma M = 0 \qquad \text{Equation 3.7}$$

Examples include a "stalemate" tug of war with no resultant movement because each team imparts an equal and opposite tensile force on the rope (Figure 3.6a and b) and a skateboarder balancing on a ramp ready to "drop in" (Figure 3.6c).

3.4 Linear momentum and impulse

Linear momentum (p) is a vector defined as the product of mass and velocity.

$$p = mu \qquad \text{Equation 3.8}$$

Figure 3.6 Examples of static equilibrium, (a and b) a "stalemate" tug of war and (c) a skateboarder balancing on the top of a ramp ready to "drop in".

For example, the momentum of a 100 kg cyclist travelling at 10 m/s would be 1,000 kg·m/s. The momentum of a 0.45 kg football travelling at the same velocity as the cyclist would be >200 times lower, at 4.5 kg·m/s, due to it being much lighter.

The change in momentum of an object is the product of its mass and its change in velocity.

$$\Delta p = m\Delta u = m\left(u_f - u_i\right) \qquad \text{Equation 3.9}$$

The change in momentum of an object is also equal to its impulse (J):

$$p_f - p_i = \Delta p = J \qquad \text{Equation 3.10}$$

A closed system is not affected by external forces, so the total momentum is conserved. This means the total momentum before an event, such as an impact, is the same as after it. If the 0.45 kg football in the previous example was initially still (u_i = 0) before being kicked to give a velocity of 10 m/s (u_f = 10 m/s), its change in momentum and hence impulse would be 4.5 kg·m/s. In this example, momentum is transferred from the foot to the football. As the football gained velocity (i.e., accelerated) from the impact the foot slowed down. Increasing the pre-impact velocity of the foot would increase the post-impact velocity of the football, and hence its change in momentum. Now imagine a golfer swinging a club to strike a stationary golf ball. They can increase the pre-impact momentum of their club, and hence the post-impact velocity of the ball, by increasing their swing speed (u), club mass (m) or both (u and m).

Considering the relationship between momentum and impulse in more detail, impulse is the product of force and time:

$$J = F\Delta t \qquad \text{Equation 3.11}$$

For example, the area under the temporal force plot in Figure 3.4b is the impulse, which can be obtained by integrating the force with respect to time:

$$J = \int_0^T F(t)\,dt \qquad \text{Equation 3.12}$$

where T is the impact duration.

Combining Equation 3.8 and Equation 3.11 gives:

$$F\Delta t = m\Delta u \qquad \text{Equation 3.13}$$

Rearranging Equation 3.13 to make force the subject gives Equation 3.1 ($F = ma$, Newton's second law). Hence, the larger the impulse of the striking object, the larger the force acting on the ball being struck, and the higher its post-impact velocity.

These concepts are often applied when testing protective devices, particularly helmets. During testing, the helmet can be fitted to a headform (i.e., artificial head surrogate) and dropped onto a force platform. While the peak force for the resulting temporal force plot (e.g., Figure 3.7a and b) gives an indication of the severity of the impact, it does not give a complete picture, as it lacks information on its duration.

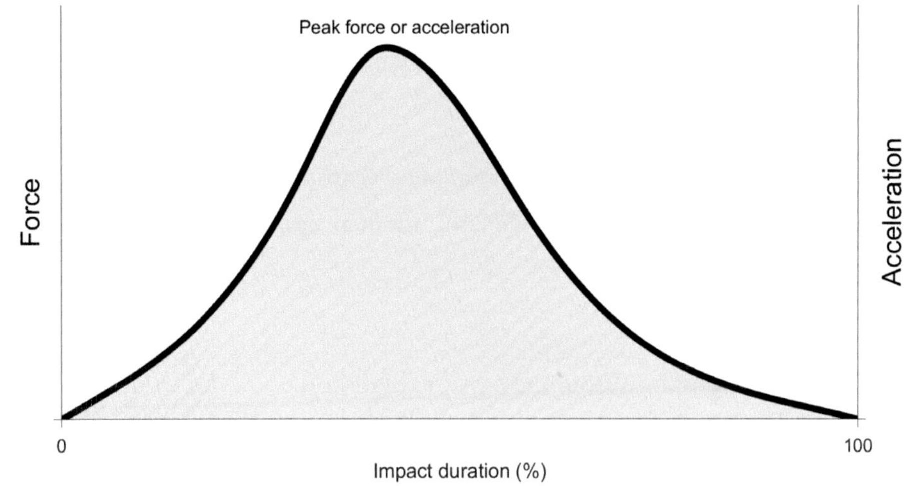

(a)

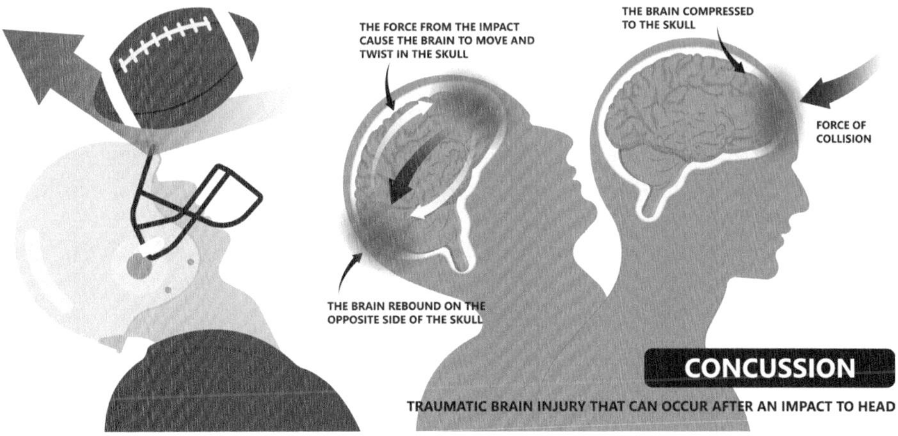

(b)

Figure 3.7 Illustrations showing (a) a typical temporal force or acceleration plot for a helmet drop test and (b) a concussion-inducing impact to the head from an American football.

The impulse of the impact can be calculated as the area under the temporal force plot (Equation 3.12), which equates to the mean force. It gives a better indication of the impact severity and the effectiveness of the helmet. The severity index (*SI*) takes this concept a step further by creating values that can be compared to reference values for head injuries to better gauge the severity of an impact and the corresponding injury risk.

$$SI = \int_0^T a(t)^{2.5}\, dt \qquad \text{Equation 3.14}$$

Note that Equation 3.14 features acceleration rather than force. Acceleration can be obtained by dividing the impact force by the combined mass of the helmet and headform (Equation 3.1). Alternatively, an accelerometer can be placed at the centre of mass of the headform to obtain its acceleration. Note the shape of the curve in Figure 3.7a is the same for force and acceleration, as the latter is simply the former divided by a constant mass. The severity index was later developed into the head injury criteria (see ref. [86]).

3.5 Angular motion, centripetal acceleration, and centripetal force

Angular motion refers to the rotation of a body about an axis of rotation. Angular velocity (ω) is defined as:

$$\omega = \frac{u}{r} \qquad \text{Equation 3.15}$$

where r is the radial distance from the axis of rotation (i.e., radius of the circle) (Figure 3.8a i). Note u is still the linear velocity of the object, despite it moving in a circular path. In this subsection, this linear velocity of the object is called tangential velocity as it is tangential to its circular path. Rearranging Equation 3.15 we see that tangential velocity is the product of angular velocity and radial distance.

Imagine a hammer thrower (Figure 3.8b) rotating with constant angular velocity. With arms extended rather than flexed, the radial distance of the hammer from the axis of rotation (their body) is greater, giving a higher tangential velocity (and hence longer flight when released).

Even when rotating at constant angular velocity the object is accelerating, which creates a force. This acceleration is due to the changing direction of the tangential velocity (Figure 3.8a i), despite it having constant magnitude. It is called centripetal acceleration, and it acts inwards towards the centre of rotation.

Centripetal acceleration will now be derived, with the help of Figure 3.8a. The angular displacement of an object moving in a circular path is defined as:

$$\theta = \frac{S}{r} \qquad \text{Equation 3.16}$$

where S is the length of the arc of rotation (Figure 3.8a ii). As the change in velocity is purely from a change in direction, we can draw a triangle (Figure 3.8a iii) and see that:

$$\theta = \frac{\Delta u}{u} \qquad \text{Equation 3.17}$$

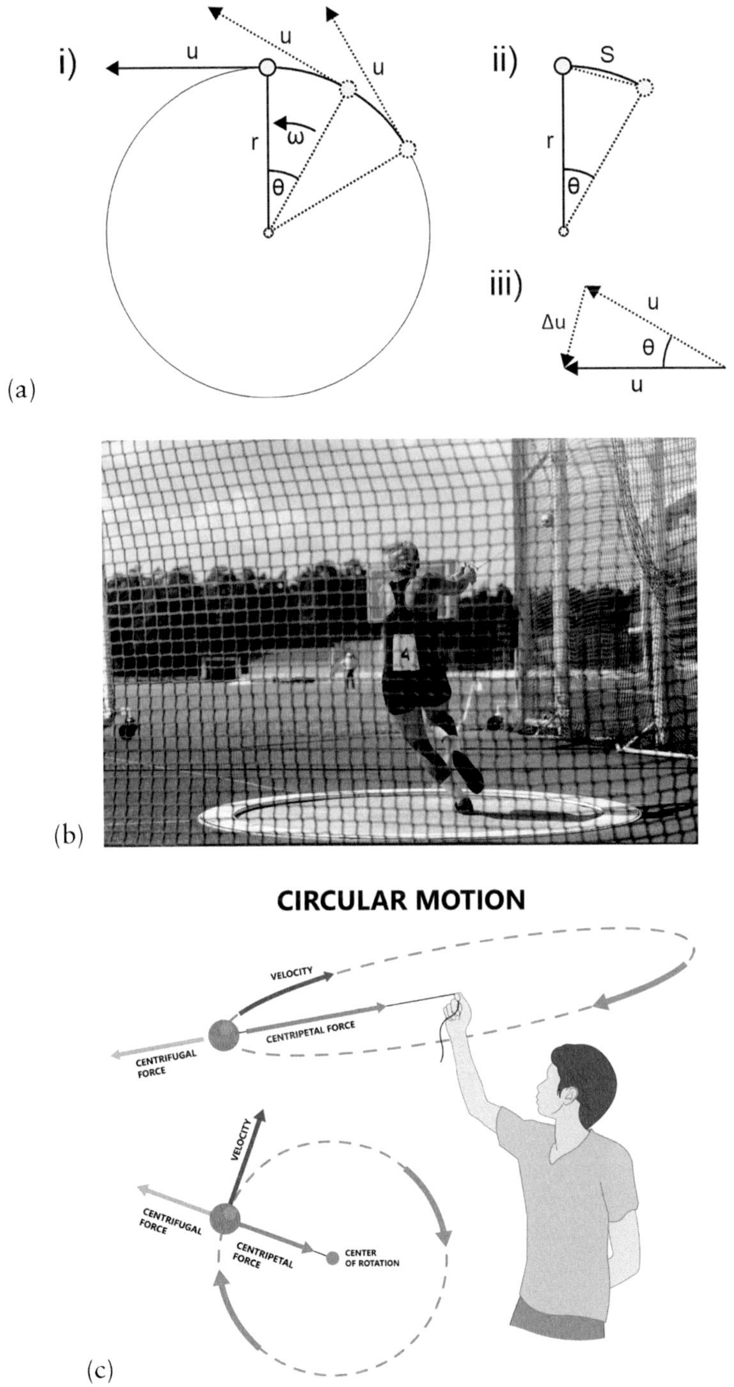

Figure 3.8 Concepts of angular motion, (a) illustration to help communicate derivation of acceleration arising from angular motion, (b) hammer thrower with arms extended to increase the radial distance from the axis of rotation, and hence the tangential velocity, and (c) illustration of forces arriving from the angular motion of a ball on a string.

For small, almost infinitesimal time periods, the angular displacement is small, and the arc of rotation can be assumed straight (i.e., chord, Figure 3.8a ii). As the angle in Equation 3.16 is the same as in Equation 3.17, they can be combined as:

$$\frac{\Delta u}{u} = \frac{S}{r}$$ Equation 3.18

Rearranging Equation 3.18 to make change in velocity the subject gives:

$$\Delta u = \frac{uS}{r}$$ Equation 3.19

Dividing both sides of Equation 3.19 by the change in time (Δt) gives:

$$\frac{\Delta u}{\Delta t} = \frac{uS}{r\Delta t}$$ Equation 3.20

which can be simplified (see Equation 3.3, and knowing that tangential velocity is the ratio of distance and time) to give centripetal acceleration (a_c):

$$a_c = \frac{u^2}{r}$$ Equation 3.21

According to Equation 3.21, if everything else is constant, doubling the radial distance from the axis of rotation will halve the centripetal acceleration, whereas doubling the tangential velocity will make it four times larger. According to Equation 3.1, centripetal force (F_c) is:

$$F_c = ma_c = m\frac{u^2}{r}$$ Equation 3.22

According to Equation 3.15, centripetal force can also be defined as:

$$F_c = m\omega^2 r$$ Equation 3.23

Returning to our hammer throw example (Figure 3.8b), the centripetal force will apply an inward tension to the hammer cable (exerted by the arms), whereas the centrifugal force is an equal opposing tensile force (Figure 3.8c). If everything else is kept constant, doubling the mass of the hammer or the radial distance will double these forces, whereas doubling the angular velocity will increase them four times. Centripetal and centrifugal forces are found in many other sporting scenarios involving angular motion, including turning a bike, skateboard, or bobsled around a banked corner (Figure 3.9).

3.6 Moment of inertia

The moment of inertia (I) of an object is:

$$I = mr^2$$ Equation 3.24

(a)

(b)

Figure 3.9 Photos showing examples of centripetal force in sport when turning (a) a mountain bike in a berm, (b) a skateboard in a bowl, and (c) a bobsled on a banked track.

(c)

Figure 3.9 (Continued)

This means the moment of inertia of an object depends on its mass and the radial distance from the axis of rotation. For example, an athlete can tuck in or outstretch their limbs to increase or decrease their moment of inertia, respectively (Figure 3.10a). Analogous to how mass is resistance to linear acceleration (Equation 3.1), moment of inertia is resistance to angular acceleration (α):

$$T = I\alpha \qquad \text{Equation 3.25}$$

where T is torque, which is analogous to moment (Equation 3.4).

Parallel axis theorem can be applied to find the moment of inertia of an object about an axis offset from the centre of mass. Moment of inertia is particularly relevant to striking implements, such as rackets, bats, and clubs, where those with higher values (about an axis in the handle) are harder for the user to accelerate during a swing. This creates a trade-off between the moment of inertia of a striking implement and the velocity at which it can be swung [91–98], with implications for pre-impact momentum and post-impact ball velocity (see discussion of Equation 3.13).

Conservation of momentum also applies to angular motion (Figure 3.10a). Imagine a figure skater tucking their limbs in tight during a spinning manoeuvre to reduce their moment of inertia (i.e., r in Equation 3.24) so they spin faster (Figure 3.10a). A pair of figure skaters in an extended position (Figure 3.10b) will not only have a higher mass than one skater, but the location of their combined centre of mass will also be further from the axis of rotation, causing them to spin slower. Regarding the design of sporting goods, skis and snowboards made from lightweight materials will have a low moment of inertia, making them easier to rotate during jumping manoeuvres such as a 360° spin.

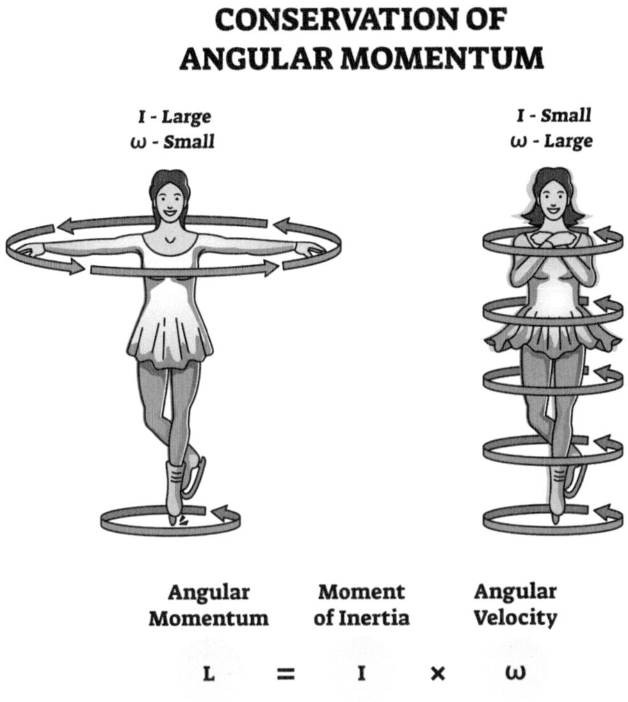

(a)

(b)

Figure 3.10 Concepts of conservation of angular momentum. Illustration (a) shows conservation of angular momentum for a rotating figure skater. With arms outstretched the skater's moment of inertia is high, and their angular velocity is low. With arms tucked in the skater's moment of inertia is low, and their angular velocity is high. Photo (b) shows a pair of figure skaters in an extended position with a high moment of inertia about their axis of rotation (i.e., the male).

3.7 Power

Power is the rate at which work is done. The units of power are watts (W). For linear motion, power is:

$$P = Fu \qquad \text{Equation 3.26}$$

For angular motion, power is:

$$P = T\omega \qquad \text{Equation 3.27}$$

The ability to generate power is key to cycling. The more power a cyclist can generate, the faster they can travel. They generate power by applying a force to their pedals via their legs. This force creates a torque about the bottom bracket via the crank arm, causing the crankset to rotate (i.e., angular velocity) and generate power. The rotation is transferred to the rear wheel via the drivetrain, propelling the bike forward.

3.8 Energy equations

There are various types of energy. Those most applicable to sport include kinetic energy, gravitational potential energy (potential energy), and strain energy. The unit of energy is the joule (J), where:

$$1\,\text{J} = 1\,\text{kg}\left(\frac{m}{s}\right)^2 = 1\frac{\text{kg}\cdot m^2}{s^2} \qquad \text{Equation 3.28}$$

The kinetic energy (*KE*) of a body is due to its motion.

$$KE = \frac{mu^2}{2} \qquad \text{Equation 3.29}$$

Returning to our 100 kg cyclist travelling at 10 m/s, their kinetic energy would be 5,000 J or 5 kJ. A runner with the same mass as the cyclist but travelling half as fast would have a quarter of the kinetic energy at 1.25 kJ.

The potential energy (*PE*) of a body is due to its position:

$$PE = mgh \qquad \text{Equation 3.30}$$

where *h* is its height from a reference point. If our 100 kg runner was to jump to a height of 1 metre above the ground (Figure 3.11a), their potential energy relative to this would be ~1 kJ. Strain energy of a body is due to its deformation (Figure 3.11b–d). The specific equations are related to the mode of deformation, as will be covered in Chapter 5.

3.9 Conservation of energy

The law of conservation of energy means the energy of an isolated system is not lost but transferred from one type to another (see Figure 3.12). For example, kinetic energy can be transferred to potential energy, and vice versa. There are many sporting examples that involve conservation of energy. Think again of our high jumper (Figure 3.11a),

(a)

(b)

Figure 3.11 Photos showing examples of conservation of energy in sport, (a) a high jumper transferring kinetic energy from the run-up to potential energy during jumping, (b) a pole vaulter transferring kinetic energy from the run-up to strain energy via bending of the pole, (c) an archer storing strain energy in the bow, which will be transferred to kinetic energy of the arrow when released, and (d) a runner storing strain energy in a running blade, which will be transferred to kinetic energy as it propels them forwards.

(c)

(d)

Figure 3.11 (Continued)

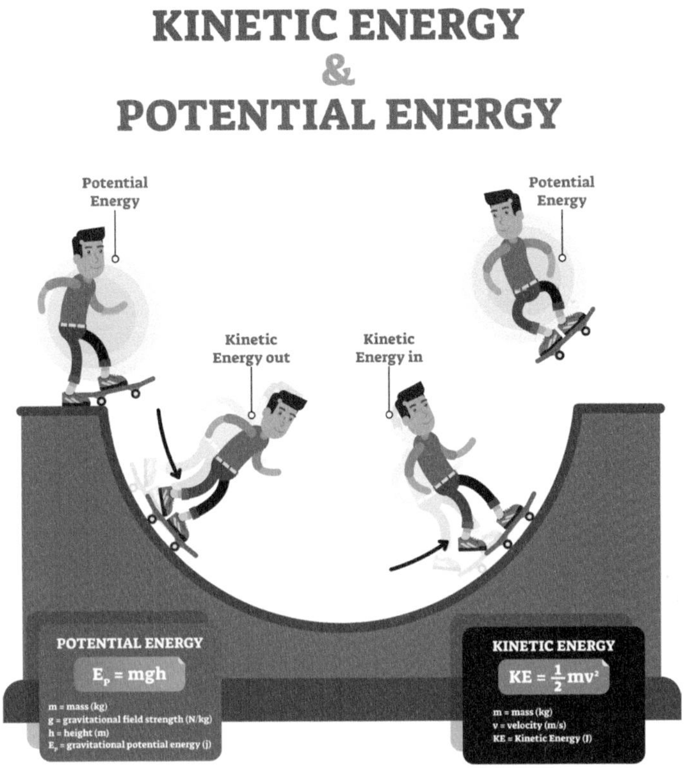

(a)

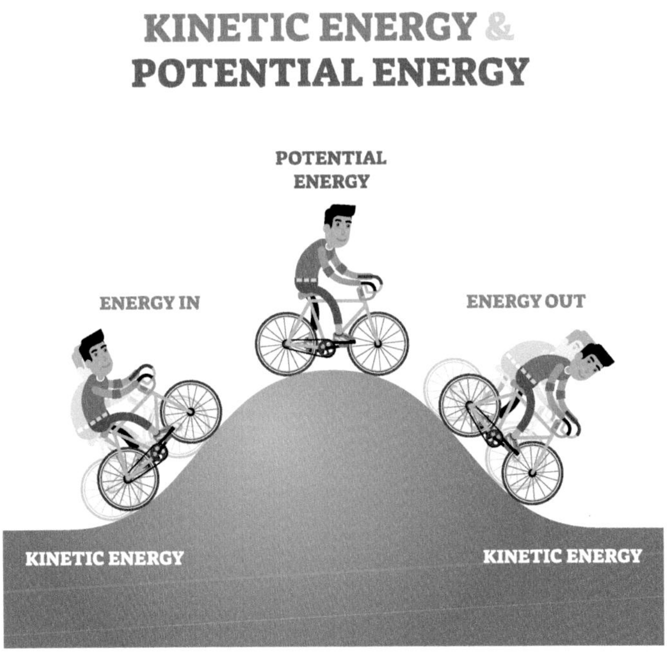

(b)

Figure 3.12 Illustrations of conservation of energy for (a) a skateboarder in a half-pipe, converting potential energy to kinetic energy and then back again, and (b) a cyclist riding over a hill, converting kinetic energy to potential energy and then back again.

who transfers kinetic energy from the run-up to potential energy during the jump. As they jump from the ground at the end of the run-up their velocity direction changes from mainly horizontal (i.e., running) to almost vertical (i.e., jumping). As they gain height during the jump and their vertical velocity slows, their kinetic energy is transferred to potential energy. Their potential energy peaks at the top of the jump, before decreasing as they return to the ground with increasing vertical velocity and hence kinetic energy.

Our high jumper in this example converts ~80% of the 1.25 kJ of kinetic energy from the run-up to ~1 kJ of potential energy at their peak height of 1 m. About 20% of their kinetic energy was retained to allow them to continue travelling forward slowly and clear the bar. Some energy would have also been transferred to other types, such as thermal and acoustic, which would not contribute to their performance. The jumper's energy is ultimately converted to strain energy as they impact the crash mat, which compresses and slows them gradually to avoid injury.

A pole vault is a more complex example of conservation of energy [99] (Figure 3.11b). In this example, strain energy is also included in the jumping (vaulting) phase, as well as kinetic and potential energy. Kinetic energy from the run-up is transferred to the pole as it bends and stores strain energy. As the pole straightens, much of the stored strain energy is transferred back to the athlete as potential energy, allowing them to gain more height than when simply jumping. Other examples of conservation of energy in sport include archery (Figure 3.11c), running on blades (Figure 3.11d), a springboard dive, and a bouncing ball.

3.10 Coefficient of restitution

Coefficient of restitution (CoR, e) relates to how safety mats perform and how balls bounce [100] (Figure 3.13). It is key to many ball sports. For a ball impacting a fixed rigid surface (e.g., when dropped on to a floor), CoR is:

$$e = \frac{u_2}{u_1} \qquad \text{Equation 3.31}$$

where the respective subscripts 1 and 2 refer to pre- and post-impact. For example, a ball rebounding at half its original velocity would have a CoR of 0.5, whereas one rebounding with three quarters of its original velocity would have a CoR of 0.75.

The proportion of kinetic energy (see Equation 3.29) retained during such a collision (collision efficiency) is:

$$\frac{KE_2}{KE_1} = \frac{\left(\frac{mu_2^2}{2}\right)}{\left(\frac{mu_1^2}{2}\right)} = \left(\frac{u_2}{u_1}\right)^2 = e^2 \qquad \text{Equation 3.32}$$

The proportion of kinetic energy transferred is therefore one minus CoR squared. A CoR of 1 corresponds to an elastic collision, with equal pre- and post-impact ball velocity (i.e., no kinetic energy transferred). A CoR of 0 corresponds to an inelastic collision, with zero post-impact ball velocity (i.e., all kinetic energy transferred). Returning to our previous example, the ball with a CoR of 0.5 retained a quarter of its pre-impact kinetic energy, whereas the one with a CoR of 0.75 retained ~56%.

The concept of CoR is often used in sporting governing body rules and regulations to control ball performance, including those for baseball and softball (ASTM F1887), tennis [101], and football [102]. In the latter two examples, when a ball is dropped from a set height, its rebound height must be within a specified range. Height is measured rather than velocity because this tends to be easier (e.g., using a video camera with a ruler positioned vertically next to the rebounding ball). The potential energy (see Equation 3.30) of the ball at the release height of such a drop test equates to its kinetic energy (see Equation 3.29) when it impacts the floor (Figure 3.13), meaning:

$$PE_1 = KE_1 \qquad \text{Equation 3.33}$$

and

$$mgh_1 = \frac{mu_1^2}{2} \qquad \text{Equation 3.34}$$

Simplifying and rearranging Equation 3.34 to make pre-impact velocity the subject gives:

$$u_1 = \sqrt{2gh_1} \qquad \text{Equation 3.35}$$

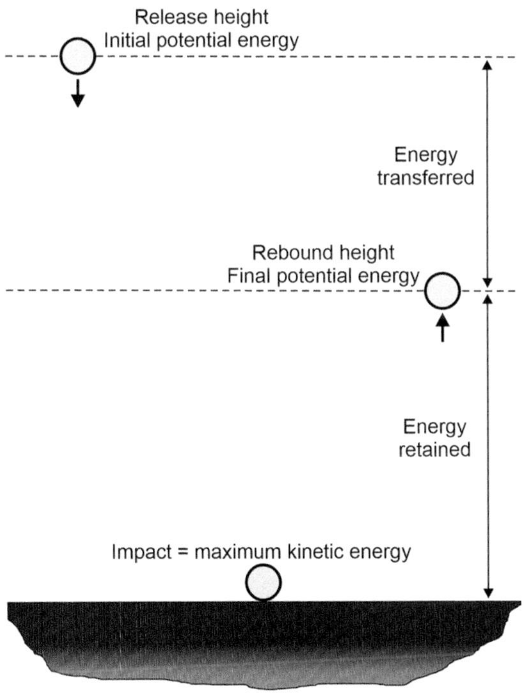

Figure 3.13 Diagram to show how conservation of energy applies to a sports ball when dropped from a set height. The ball has maximum potential energy at the release height, which is converted to kinetic energy as it falls under gravity until it impacts the floor. The ball rebounds to a height lower than the release height because some energy is transferred.

The same process can be applied to obtain the post-impact ball velocity from its rebound height, and hence:

$$e = \sqrt{\frac{h_2}{h_1}} = \frac{u_2}{u_1}$$ Equation 3.36

This also means the proportion of potential energy (see Equation 3.29) retained during such a collision is defined as:

$$e^2 = \frac{h_2}{h_1}$$ Equation 3.37

A ball rebounding to half its original height would have a CoR of 0.71, meaning it would have retained half its pre-impact potential energy. A ball rebounding to three quarters of its original height would have a CoR of 0.87, meaning it would have retained three quarters of its pre-impact potential energy.

Table 3.2 shows CoR governing body requirement examples for tennis [101] and footballs [102]. The required CoR was calculated using Equation 3.37 based on the drop and rebound heights stated in the corresponding regulations. Note that the footballs have a higher CoR requirement than the tennis balls despite the similar rebound height because their drop height is lower.

Now consider a tennis ball dropped onto the string bed of a tennis racket with its frame fixed (so it cannot recoil during an impact). Compared to an equivalent impact with the floor, the ball will rebound higher with more of its potential energy retained, equating to a larger CoR. This is because strain energy is now transferred to both the strings and the ball, and the former has a higher collision efficiency. Reducing the tension of the strings will cause the more efficient string bed to deform more and store a greater proportion of the strain energy during the collision than the less efficient ball. This means the ball will rebound faster from a tennis racket with lower string tension. This concept is known as the "trampoline effect" [92, 103] and it is applied to the design of hollow implements like metal baseball bats and golf drivers, to increase collision efficiency and hence post-impact ball speed.

Table 3.2 Examples of ball CoR requirements for governing body approval tests [101, 102]

Sport	*Governing body*	*Ball type*	*Drop height (m)*	*Required rebound height (m)*	*Required CoR range*
Tennis	International Tennis Federation (ITF)	Type 2	2.54	1.35–1.47	0.73–0.76
Football	Fédération Internationale de Football Association (FIFA)	Quality Pro	2.00	1.35–1.55	0.82–0.88

3.11 Friction

The laws of friction dictate the friction forces acting on an object on a plane. In the simplest scenario the plane is horizontal (Figure 3.14a), and the force (F_l) required to initiate in-plane movement of the object is:

$$F_l = \mu_s N \qquad \text{Equation 3.38}$$

where μ_s is the limiting static coefficient of friction (CoF) acting between the contacting surfaces and N is the normal force (weight of the object). For a given limiting static CoF, the force required to initiate in-plane movement increases with the object's weight (i.e., larger N). This force is independent of the contact area between the two surfaces. The limiting static CoF depends on the properties of the surfaces in contact. When the object moves, Equation 3.38 becomes:

$$F_k = \mu_k N \qquad \text{Equation 3.39}$$

where the subscript k is kinetic (i.e., dynamic). The kinetic CoF tends to be lower than its static counterpart (Figure 3.14b). These coefficients of friction can also be reduced by manipulating the surface. For example, skateboarders and snowboarders rub wax on rails and ledges to reduce friction forces when grinding or sliding (Figure 3.15).

Friction is applicable to many sports, although in different ways. Climbers and trail runners want high shoe–surface friction coefficients to avoid slipping (see ref. [104]), whereas ice skaters and bobsledders need low equipment–surface friction coefficients to glide. Unsurprising, rubber, as typically used on shoe soles, has high friction coefficients on rock/tarmac surfaces, whereas steel, as used for ice skate blades and bobsled runners, has low coefficients of friction on ice (Table 3.3). Regelation is the process of ice melting under pressure and then refreezing when the pressure is reduced. This process reduces ice-blade friction and helps ice skaters glide. When wheels are used (see Figure 3.14a), such as for skateboards, rollerblades, and bikes, the coefficient of friction can be replaced by the rolling resistance coefficient (RRC).

Friction typically refers to the interaction between smooth surfaces, such as a ski base and snow. When referring to footwear, particularly shoes with studs or tread patterns, the term "traction" is often used instead of friction. In this context, the footwear–surface interaction, particularly the penetration depth, can depend on the normal force and surface conditions (i.e., dry vs. mud). As such, surface traction relationships can be more complex than those described in the proceeding equations for friction. For example, unlike the CoF, the coefficient of traction (CoT) may not be a constant [105, 106].

Many of the sports where friction is a key component do not take place on a horizontal plane. In these cases, we need to consider friction on an inclined plane (Figure 3.16a), and trigonometry can be used to resolve the force vectors. The forces perpendicular (N_i) and parallel (F_i) to an inclined plane are:

$$N_i = mg\cos\theta \qquad \text{Equation 3.40}$$

$$F_i = mg\sin\theta \qquad \text{Equation 3.41}$$

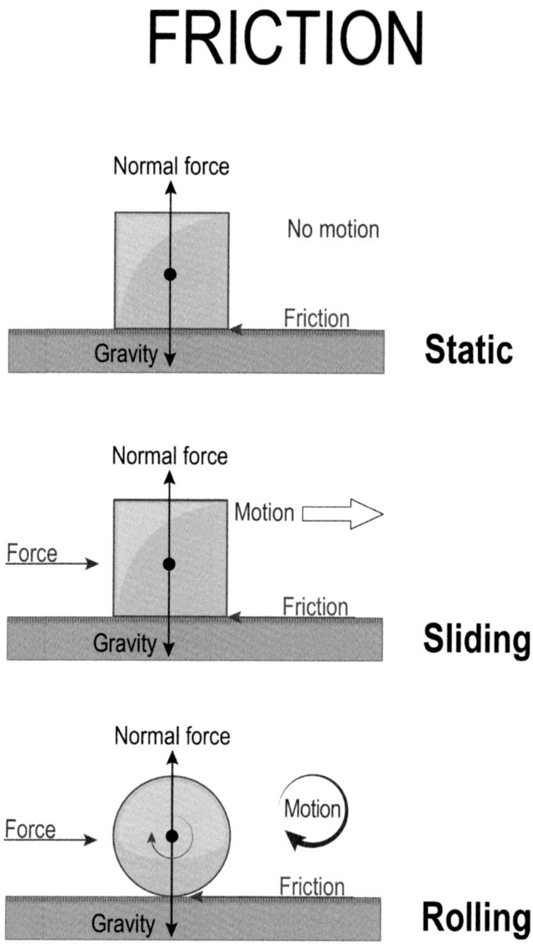

(a)

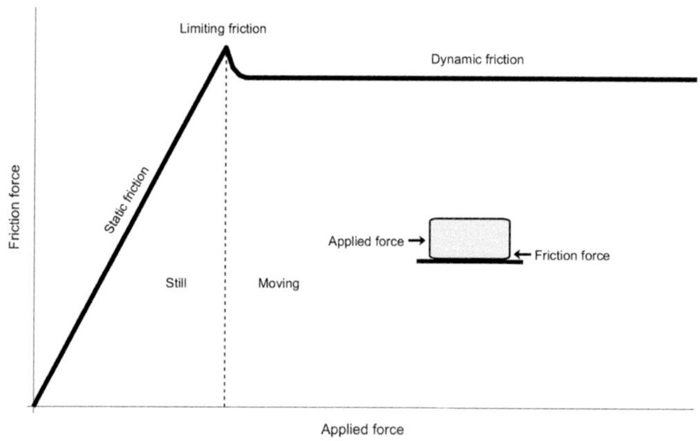

(b)

Figure 3.14 Concepts of friction for an object on a horizontal plane, (a) forces acting on the object, including friction, (b) graph showing a typical relationship between applied force and friction force. The friction force opposes the applied force, and they are equal when the object is still. The limiting friction corresponds to the point at which the object starts moving. When the object is moving, the friction force is almost constant.

Figure 3.15 Photo of a snowboarder overcoming friction forces to slide along a rail.

Table 3.3 Typical coefficient of friction values for some common sporting materials. These are just indicative, and if exact values are required, these should be obtained, either from the literature or via testing

Interacting surfaces	*Limiting CoF*	*Kinetic CoF*	*Typical examples*
Ice–steel	0.03		Bobsled, ice-skating
Rubber–asphalt/concrete	0.9	0.5–0.8	Running shoes
Waxed ski–snow		0.04–0.2	Skiing, snowboarding
Wood–concrete	0.6		Skateboard slide

Equation 3.38 and Equation 3.39 can then be applied in the same manner, as required. A classic sporting example is a skier on a ski slope (Figure 3.16b). As the slope steepens, the ratio of parallel to perpendicular force increases, and the skier slides downhill more easily.

Friction also applies to ball sports [107]. When a ball impacts a surface at an oblique angle (i.e., not perpendicular) the friction force at the contacting region produces a torque about its centre of mass, inducing spin. Tennis is an example where players impart topspin on the ball by angling the face of their racket. This spin influences the flight of the ball, as discussed in Chapter 4, which covers fluid dynamics in sport.

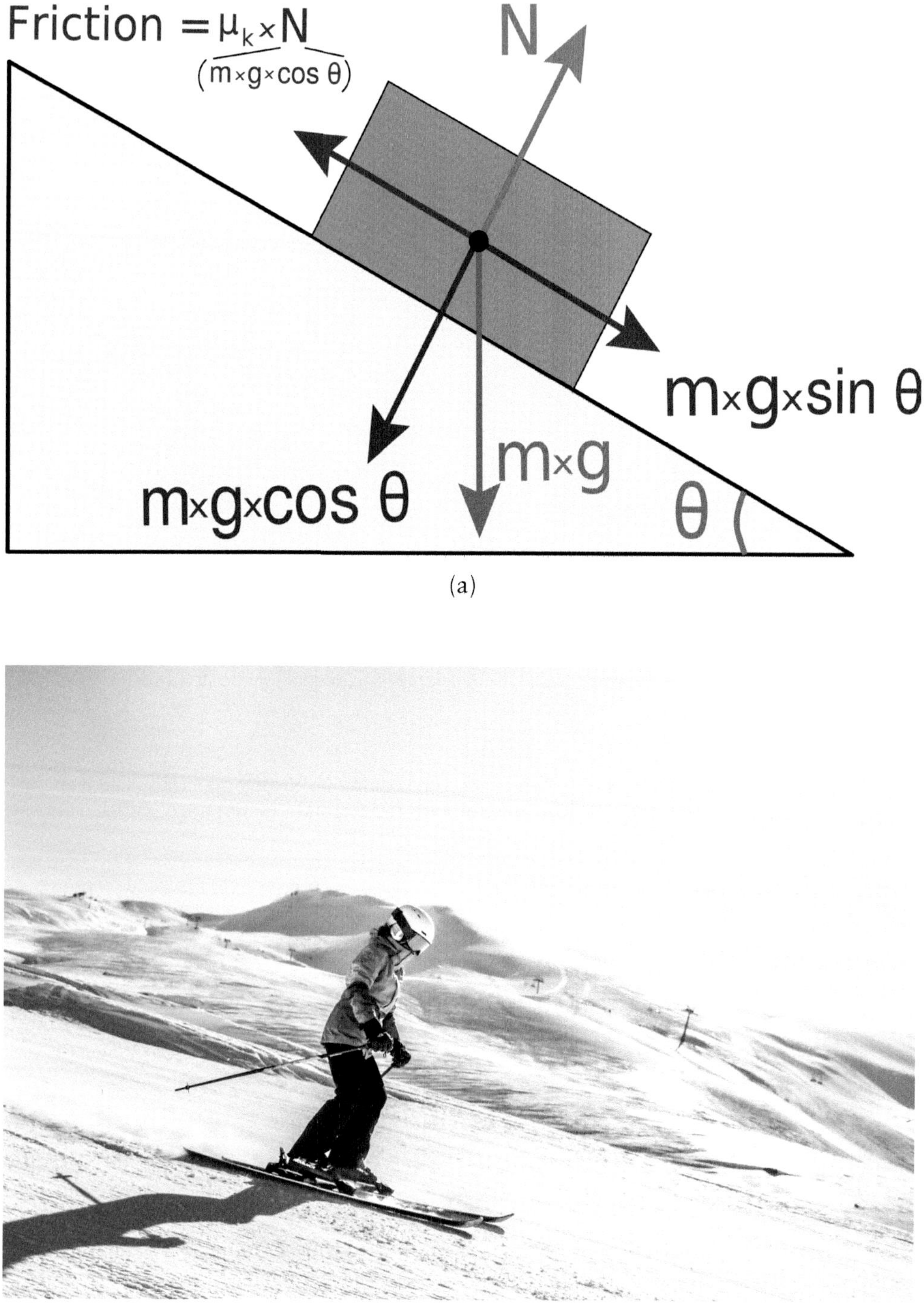

(a)

(b)

Figure 3.16 Concepts of friction for an object on an inclined plane, (a) illustration of friction for an object on an inclined plane, and (b) a skier overcoming ski-snow friction as they slide down a piste.

3.12 Chapter summary

This chapter has provided a broad overview of sports mechanics. It began with an overview of basic mechanics concepts and their application in sporting scenarios. References were also made to how these concepts apply to the design of sporting goods. Readers should now have a basic understanding of the role and importance of mechanics to sports engineers. This basic understanding should be sufficient to see how mechanics relates to the complementary topics covered in this introductory textbook on sports engineering. However, there is still more to be learnt about mechanics beyond the basics covered here. Students are encouraged to continue reading and learning about mechanics and how it applies to sport.

3.13 Suggested further reading

The following suggestions for further reading go deeper into the topics introduced in this chapter:

- *Gold Medal Physics: The Science of Sports* [108]
- *Routledge Handbook of Sports Technology and Engineering*, 1st and 2nd editions [10, 86]
- *Gliding for Gold: The Physics of Winter Sports* [109]
- *Physics of Baseball and Softball* [92]

4 Fluid dynamics in sport

Fluid dynamics plays an important role in sport and for this reason the subject has been a focus of study by sports engineers for decades. When we discuss fluid dynamics in sport, what we are predominantly interested in are the sub-disciplines of either aerodynamics (interactions between gases and solid bodies) or hydrodynamics (interactions between liquids and solid bodies).

Sports engineers attempt to control and manipulate the aero- and hydrodynamic characteristics of the sporting goods they develop, as these can influence the performance of the item or outcome of a competition in subtle or dramatic ways. It is through understanding sports fluid dynamics that sports engineers may identify subtle alterations to the position of seams on an athlete's racing suit, which may result in a medal. It is the same flow physics that can dramatically influence the visible motion of sports equipment, e.g., the swerve in flight of an association football, which may or may not be an intentional aspect of its design!

This chapter covers the fundamentals of sports fluid dynamics, though with emphasis on sports aerodynamics. Basic fluid dynamic properties and concepts are discussed before moving into more detailed topics such as aerodynamic forces in sport, their impact on equipment behaviour, and how we can control these to our advantage.

4.1 Why are fluid dynamics in sport important?

Within sport we are interested in reducing, controlling, or harnessing the resulting fluid dynamic forces that are created when a fluid, be it gas or liquid, moves over, around, or through an object. Hydrodynamic examples include the reduction of resistive force a swimmer experiences through development of swimsuit technologies [50, 110, 111], controlling water movement around the fins and hull in all manner of waterborne craft, or optimising rowing oar designs to harness and maximise propulsive motion. Aerodynamic examples include controlling the forces that cause a football free kick to swerve beyond an outstretched goalkeeper, as in Figure 4.1a, or that can extend the flight time of a golf ball, through manipulation of its shape and textures, as in Figure 4.1b. Within Formula 1, race cars only remain firmly on the ground whilst cornering at extreme speeds by harnessing aerodynamic downforce, as in Figure 4.1c. Resistive force components also play a key role in competitive performance and outcomes.

DOI: 10.4324/9781003411086-4

(a)

(b)

Figure 4.1 Photos showing (a) football free kick, (b) golf ball in flight, and (c) Formula 1 cars at high speed.

(c)

Figure 4.1 (Continued)

A classic example of the influence of fluid dynamics upon competitive performance and race outcome was seen in the 1989 Tour de France cycling race final stage. This edition of the Tour had been a close race dominated by two competitors, the Frenchman Laurent Fignon and the American Greg Lemond, who had traded the yellow jersey over the preceding stages. With 3,242 km of gruelling racing completed, only a single-stage 25 km time trial into Paris remained. It was Fignon who went into this final stage with a 53 second lead over his rival, a gap many believed would be insurmountable over such a short distance. However, throughout the previous time trial stages of the 1989 Tour, Lemond had adopted the use of aerodynamic tribars, previously only used in triathlons, and had worn an "aero helmet". This had the effect of streamlining his position upon the bicycle, reducing the aerodynamic forces that could slow him down. Fignon did not use any such equipment. With his aerodynamic advantage Lemond had finished ahead of Fignon in each preceding time trial. The final was to be no different, for it was the American who would triumph, not only overhauling the Frenchman's 53 second lead but finishing a further 8 seconds clear to win the Tour. It has been estimated that Lemond's aerodynamic advantage may have been closer to 90 seconds, demonstrating despite the immense effort Fignon clearly exerted to attempt to win the race, Lemond's equipment choice was a key factor in his victory. From then onwards the use of aerodynamically designed equipment in cycling grew and is now commonplace in racing (see Figure 4.2).

Cycling is not the only sport where aerodynamic performance has an influence on race outcome. Examples of other sports where aerodynamics is an important consideration of the resistive forces encountered include the winter sport of speed skating (up to 90% of total resistive forces) [112], alpine skiing (~80%) [113], and bobsled (~25%) [114]. Runners also experience aerodynamic resistance, but to a lesser extent, accounting for only 7~9% [115] of their resistive forces. For this reason, aerodynamics is often overlooked in running. However, its importance was again seen when Eliud Kipchoge (unofficially) broke the two-hour time record in October 2019 for a marathon in a specially staged event, the Ineos 1:59 Challenge held in Vienna, Austria. Kipchoge utilised pace setters

Figure 4.2 Photo showing a cyclist using an aerodynamic helmet and triathlon bars.

carefully positioned around him in a set formation to reduce the aerodynamic resistance forces he experienced.

4.2 Density, pressure, and viscosity

To understand how sport is affected by fluid dynamics and how we can control and reduce resistive fluid forces, we first need to consider some basic properties and principles. We will cover some basics within this section, but do not go into lengthy dialogue or explanation. There are plenty of classical fluid dynamics and mechanics textbooks available to the reader, such as White [116] and Massey and Ward-Smith [117], for full derivations and detailed explanation.

4.2.1 *Pressure*

All fluids have an associated pressure measured in pascals (Pa). The intensity of that pressure at a point is defined as the force acting normal to the unit area (A) of an object, and is defined as:

$$P = \frac{F}{A} \qquad \text{Equation 4.1}$$

All objects have an atmospheric pressure acting upon them. They also experience local changes in gauge pressure associated with fluid movement, as will be discussed later. The combination of these two pressures is termed the absolute pressure.

$$P_{Absolute} = P_{Gauge} + P_{Atmospheric} \qquad \text{Equation 4.2}$$

4.2.2 Density

In sport we are primarily only ever interested in two fluids and their respective properties, water and air. These two fluids are notably different due to their respective fluid density (ρ), the mass (m) of a substance per unit volume (V).

$$\rho = \frac{m}{V} \qquad \text{Equation 4.3}$$

At a standard temperature and pressure of T = 15°C and P = 101325 Pa, the density of air is ρ_{air} = 1.225 kg/m^3, whereas water is over 830 times denser ρ_{water} = 1,000 kg/m^3. The density of both fluids is inversely affected by temperature. Fluid density will decrease as temperature increases, and vice versa.

The density of a fluid is also affected by pressure in a relationship that is directly proportional. Density increases as pressure increases, and vice versa. However, our two fluids differ in how they can be compressed by the pressure they experience, the force applied per unit area (see Equation 4.1).

Water is essentially incompressible, with negligible change in its density for any change in pressure. In contrast, air is compressible, with notable changes in its density associated with variation in pressure. We can calculate the influence of pressure and temperature on air density using:

$$\rho = \frac{P}{R_{specific} T} \qquad \text{Equation 4.4}$$

where $R_{specific}$ is the universal gas constant for air, $R_{specific}$ = 287 J/KgK. It is also possible to refer to published tables and graphs of fluid properties to determine the density of a specific fluid [118].

It is important to know the fluid density when calculating aerodynamic forces. Changes in the fluid density will directly influence the forces calculated or measured. For example, aerodynamic resistive forces will be affected by whether a sport is played in a cold or hot climate, and at sea level or at altitude. An increase in temperature or altitude will cause air density to fall. As a result, the aerodynamic resistive forces experienced by an object will also decrease. Therefore, a kicked ball will travel further at high altitude, and a cyclist would travel faster for the same amount of energy expended. It is for this reason that several cyclists have attempted to break the hour record for distance covered in velodromes at altitude, e.g., in Mexico City located 2,300 m above sea level.

4.2.3 Viscosity

A fluid's viscosity describes its ability to resist deformation when subjected to an applied shear stress.

$$\tau = \mu \frac{\delta_u}{\delta_y} \qquad \text{Equation 4.5}$$

All "real" fluids possess a viscosity. To explain this think of a fluid as consisting of a series of stacked layers that move with and in relation to each other. If you applied a

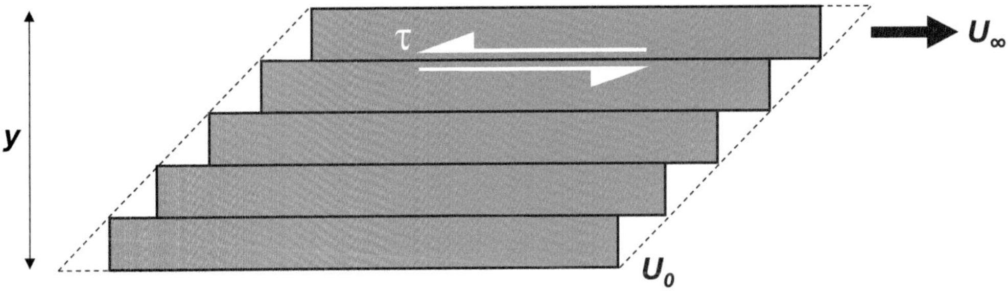

Figure 4.3 Diagram showing shear stresses within fluid layers subject to a velocity gradient.

shear stress to one of these layers, its adjoining layers would attempt to resist that motion, with the resistive forces acting aligned in the opposing direction. The resistance is only evident when fluid movement is occurring. However, any displaced layers will remain displaced when the shear force is removed.

Our two fluids have very different viscosities. Water has a viscosity of $\mu_{water} = 1.1375 \times 10^{-4}$ Pa.s and is classically termed a "thin" liquid, as it can be readily poured, unlike a "thick" liquid such as honey. Air also has a viscosity $\mu_{air} = 1.802 \times 10^{-5}$ Pa.s, and although this is much lower than for water and perhaps less evident as we do not "pour" air, it is still important.

This is because as a fluid moves over the surface of an object, a velocity gradient will be formed between that surface where the fluid will come to rest and the fluid flowing above. The region that this velocity gradient adjacent to the surface resides in is termed a boundary layer and is important to the subsequent forces measured on that object as will be shown.

The above term μ is referred to as the dynamic viscosity of a fluid and is the term we predominantly refer to. However, viscosity is also presented in literature in its kinematic form υ, which accounts for the fluid's density.

$$\upsilon = \frac{\mu}{\rho} \qquad \text{Equation 4.6}$$

4.3 Bernoulli's principle

Sport is not a static event! Whether the subject of study is a ball in flight, a cyclist, or a runner, the pressure acting on it will vary depending on its velocity and shape. To understand the forces that act on that object, we need to understand the pressure acting on its surface. A simple starting point is to calculate the dynamic pressure acting on the object through use of Bernoulli's principle.

Daniel Bernoulli described the principle in his 1738 book *Hydrodynamica* [119], which states that due to the principle of conservation of energy (see Chapter 3), the total mechanical energy at a point on a streamline within a steady fluid flow must remain constant at all points along it. Put simply, this means if you follow a particle of fluid along its flow path, as its velocity (u) increases its pressure (P) will be observed to fall, and vice versa. This principle written in full gives the Bernoulli equation:

$$\frac{P_1}{\rho} + \frac{1}{2}u_1^2 + gh_1 = \frac{P_2}{\rho} + \frac{1}{2}u_2^2 + gh_2 \qquad \text{Equation 4.7}$$

The fluids we deal with in Sports Engineering are typically classed as being incompressible, the Mach number associated with them being less than 0.3 times the speed of sound. Mach number M is the ratio of the local flow velocity and the speed of sound in the fluid (c):

$$M = u / c \qquad \text{Equation 4.8}$$

Additionally, we assume that changes in height around the object have a negligible effect, so Equation 4.7 simplifies to:

$$P + \frac{1}{2}u^2 = const \qquad \text{Equation 4.9}$$

This simplified equation includes the local static pressure (P) of the flow, and the dynamic pressure ($\frac{1}{2}u^2$) due to flow velocity.

There is, however, a problem with the Bernoulli equation. It is only valid for flows that are, or near to being, inviscid, flows that have no viscosity. As already discussed, the flows that we are interested in understanding are "real" with viscous affects. Consider movement of fluid around a ball as shown in Figure 4.4. If the flow was inviscid it would move in a smooth and balanced manner around the ball, meaning that there would be no net force acting on the ball! Real flows contain viscous effects such as flow separations and aerodynamic wakes, which swirl in an unpredictable manner. These flow phenomena then create the aerodynamic forces on the ball. This inviscid limitation does not mean, however, that Bernoulli's equation is of no use. Although simple in form, it is a useful starting point for understanding what is going on and what to expect.

For example, we can determine the stagnation pressure acting on the leading surface of an object. Air striking the front of the ball will stagnate; it will decelerate to rest at this point (Figure 4.5). We term the pressure at this point as being the stagnation pressure, the pressure attributable to the deceleration from freestream velocity to zero. The stagnation pressure as determined from summing the static and dynamic components in Equation 4.9 is useful as a starting point for understanding the magnitude of force the ball may experience. This can also help us validate, to an extent, any experimental or computational analysis of the flow around the ball.

4.4 Reynolds number

The influence of viscosity that controls forces experienced by objects within a flow is complex. For real fluids flowing around an object, we have two basic classifications of that movement, laminar or turbulent. Laminar flows are typically slow-moving with streamlines that flow around an object in an organised, often aligned, manner. Turbulent flows by contrast are characterised by unsteady and irregular motions fluctuating in time and scale. The point at which a flow switches from a laminar to turbulent behaviour is dependent on many factors. These factors include both the properties of the fluid and the shape and texture of the object it is flowing around.

We can determine whether the motion within a fluid is classified as being laminar or turbulent due to the work of Osborne Reynolds who, in 1883, proposed a relationship that determined the ratio of inertial to viscous forces in a flow. In a laminar flow, viscous forces will be dominant, keeping streamline movements inline, whereas turbulent flow is

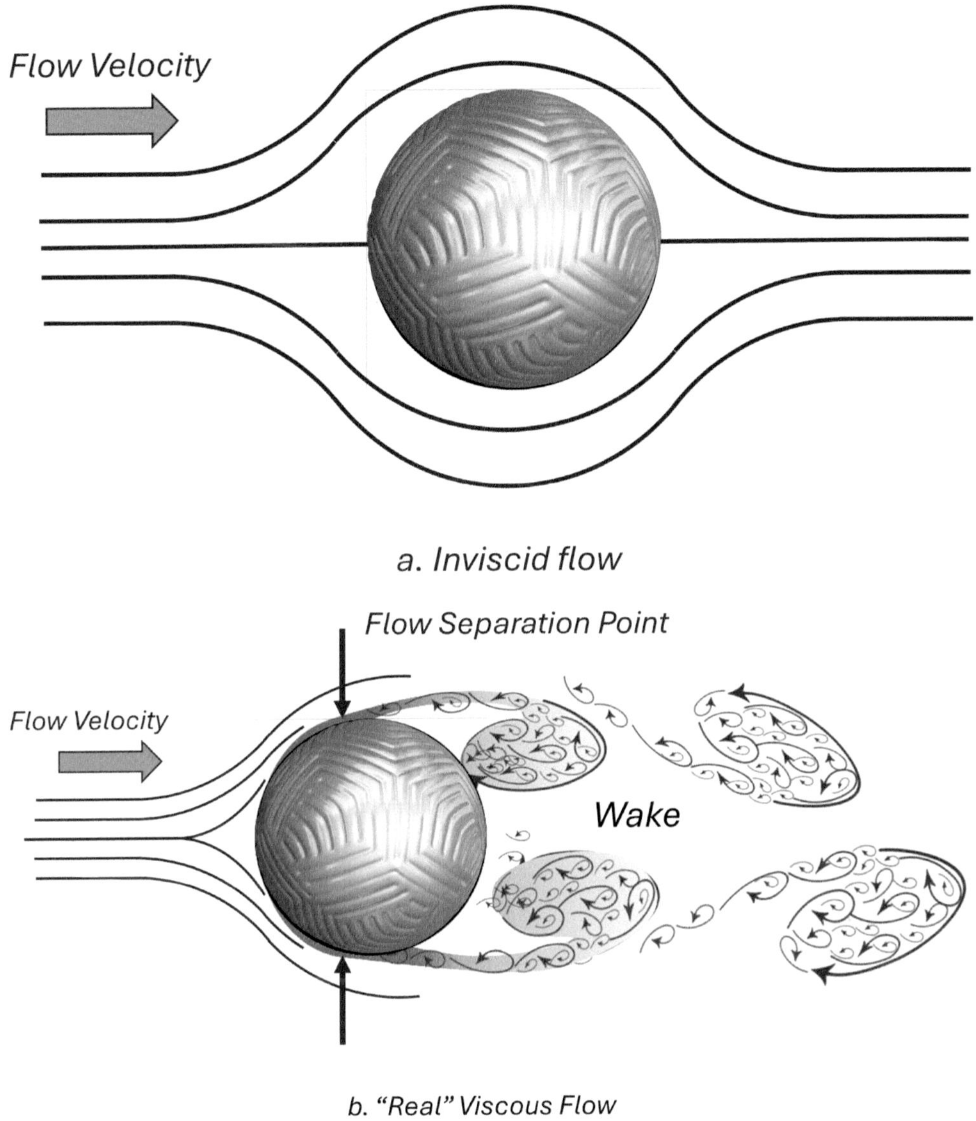

Figure 4.4 (a) Illustration of inviscid flow around a ball, where there are no viscous effects and streamlines move around the ball in a smooth, balanced manner meaning there can be no forces acting on the ball, and (b) illustration of real viscous flow moving around a ball, where viscous effects such as flow separation and subsequent wake structures create aerodynamic forces acting on the ball.

dominated by inertial forces and motion becomes unpredictable. Reynolds number (Re) is non-dimensional.

$$Re = \frac{\rho u D}{\mu} \qquad \text{Equation 4.10}$$

This is determined from the freestream velocity of the flow, the density and the viscosity, but also requires a characteristic length (D).

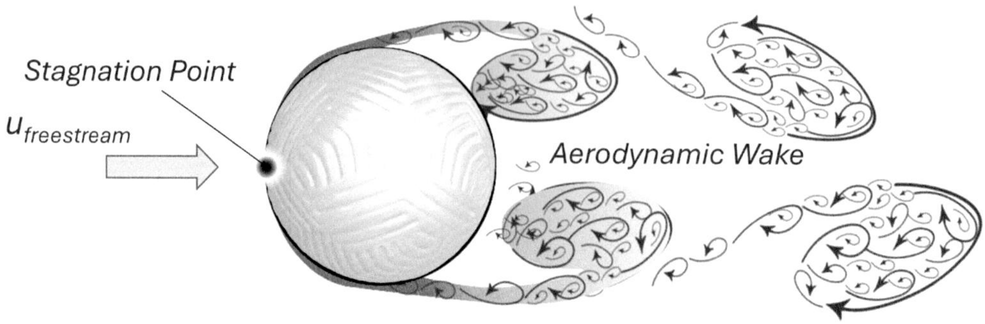

Figure 4.5 Illustration of real flow around a football in flight, showing the aerodynamic wake that forms behind the ball due to viscous effects, and the location of the flow stagnation on the front of the ball where freestream flow decelerates to a rest on the ball's surface.

Table 4.1 Ball sport comparisons of associated balls' diameters, typical velocity in play, and calculated Reynolds number

Sport	*Diameter (m)*	*Velocity (m/s)*	*Re*
Table tennis	0.038	30	77,500
Squash	0.041	30	83,600
Cricket	0.073	30	148,900
Golf	0.043	70	204,600
Tennis	0.066	50	224,300
Football	0.218	25	370,500

The characteristic length is determined in sports applications typically as the largest dimension associated with the object of interest, e.g., the diameter of a sports ball. Typical values for sports balls are shown in Table 4.1. Although the tabulated velocity for a cricket ball and squash ball is identical, differences in their diameters produces different Reynolds numbers. Whereas for the examples of golf and tennis, although their diameters and velocities differ, they return *Re* of comparable magnitude.

Reynolds determined that conditions within a flow could be considered similar if the value of *Re* were similar. This is of use in not only the comparison of objects of different sizes but also in the scaling of wind tunnel experiments. For example, using the values in Table 4.1, if you did not have access to a wind tunnel large enough to accommodate a full-sized football, you could use a scale model. If the football was scaled to 50%, you could still obtain Re = 370,500 flow conditions if your tunnel was able to run up to a velocity of 50 m/s (i.e., double).

4.4.1 Boundary layers

Because of the effects of viscosity, any fluid flowing around an object will have zero velocity at its surface. This is because real fluids moving over a solid surface cannot slip. Therefore, a velocity gradient will exist between the surface of an object where $u = 0$ and the freestream of the flow, u_∞. This surface-adjacent region of velocity gradient is termed the boundary layer.

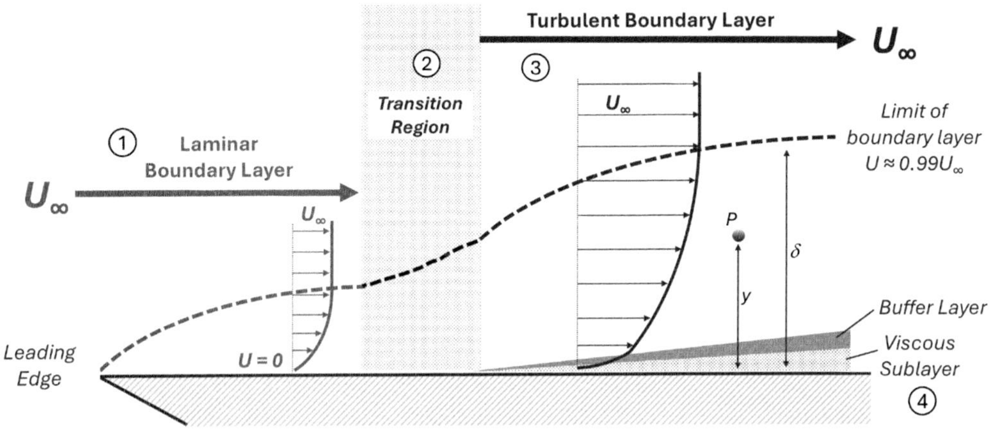

Figure 4.6 Illustration of basic boundary layer growth from the leading edge of a solid surface. The initial laminar layer (1) grows in thickness and instability transitioning (2) to a fully turbulent boundary layer (3) that grows rapidly in thickness. The rapidly thickening turbulent boundary layer develops a high velocity gradient close to the wall surface (4) a viscous sublayer that essentially behaves as a laminar layer.

A simple boundary layer forming from the leading edge of a flat plate is shown in Figure 4.6. This boundary layer is initially laminar in nature, moving in a smooth orderly manner with little mixing occurring. If the flow is moving over a smooth relatively short flat surface, and remains at a low velocity, this boundary will remain laminar and thin. However, such scenarios are rare, and if the flow accelerates or the wall roughness increases, the laminar layer will increase in both thickness and instability. This instability results in the layer transitioning into a turbulent boundary layer, rapidly over a short distance. The growth of the turbulent boundary layer then accelerates due to an increase in fluid mixing within the layer. A more uniform velocity profile exists across the layer due to the increased mixing, and as flow must still come to rest at the wall surface, a steep velocity gradient adjacent to the wall forms. This region of steep gradient essentially acts as a laminar layer and is termed the viscous sublayer; it blends into the fully turbulent region through a zone termed the buffer layer.

The rate at which a boundary layer transitions and grows is not only dependent on the speed of the flow, smoothness, and length of the surface but also on the shape of that surface. Figure 4.7 shows an example of a boundary layer forming and moving over a curved surface, e.g., a ball.

When the flow first contacts the ball, an initially laminar boundary layer is formed, close to the stagnation point on the front of the ball. The layer will begin to grow (thicken) as it moves up over the ball's curved surface. As the boundary layer moves over the ball, it will accelerate, and it may transition into a turbulent boundary layer. This could be attributable to this local flow acceleration, or the roughness or texture of the surface it is moving over. Regardless, as the boundary layer reaches the apex of the ball surface, its velocity will have reached a maximum. As we have already seen due to the principle of conservation of energy, if flow accelerates, the pressure falls and is therefore at a minimum at this point. From the stagnation point on the frontal surface of the ball up to this apex, the pressure gradient is negative as the flow accelerates. However, beyond the apex,

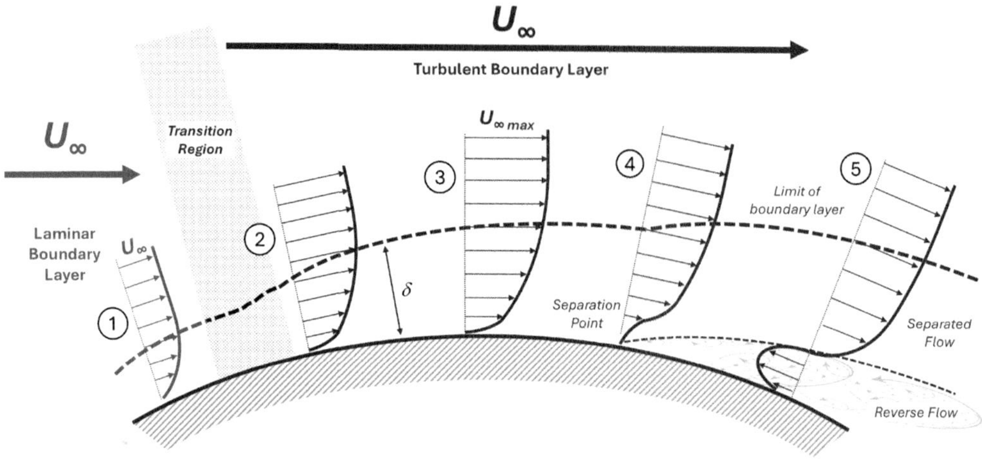

Figure 4.7 A simplified illustration of boundary layer growth over a curved surface. An initial laminar boundary layer (1) that forms from the stagnation point of the object's leading edge accelerates over the object's surface, transitioning (2) to a turbulent state. Flow continues to accelerate over the surface until at the apex (3) flow reaches maximum velocity and minimum pressure. As the boundary layer moves beyond the apex, pressure increases, creating an adverse pressure gradient that will lead to further deceleration of the boundary layer adjacent to the surface (4) that will lead to a flow separation. Once the flow has separated from the surface, a wake of separated flow is formed and the void between the wake and the object's surface is filled by reversed flow movement.

pressure will recover, increase, and begin to oppose the forward motion of the boundary layer flow. As the flow velocity adjacent to the surface already is reduced to zero, this increase in pressure acts to further slow the boundary layer. This slowing is not only adjacent to the ball's surface but into the layer itself, which can cause flow to separate from the ball. At this point a reverse in flow direction occurs, as flow reverses into the void created by the separated flow. The separated flow forms a wake behind the object.

It is important that we understand boundary layers as fluid forces acting on an object, such as drag, which are dependent upon the behaviour of the boundary layer. As we have already stated, the behaviour of the boundary layer will be influenced not only by flow velocity, but also by the shape, roughness, and texture of the object's surface. We will discuss later in this chapter how we can utilise surface roughness or texture to manipulate the flow behaviours of boundary layers to our advantage, controlling the points of flow transition and influencing any subsequent flow separation. As will be shown, a ball's texture is key to its behaviour in play.

4.5 Drag force

Flow moving over an object is brought to rest at the surface of that object by the action of viscous force. This resistive viscous force acting tangentially at the surface against the movement of the flow is termed *viscous drag*, or more commonly as *skin friction drag*. If the freestream flow is aligned parallel to the object's surface, the only force acting upon that surface will be this frictional force. However, when the freestream is no longer parallel to the surface of an object, an additional drag comes into play, *pressure* or *form drag*.

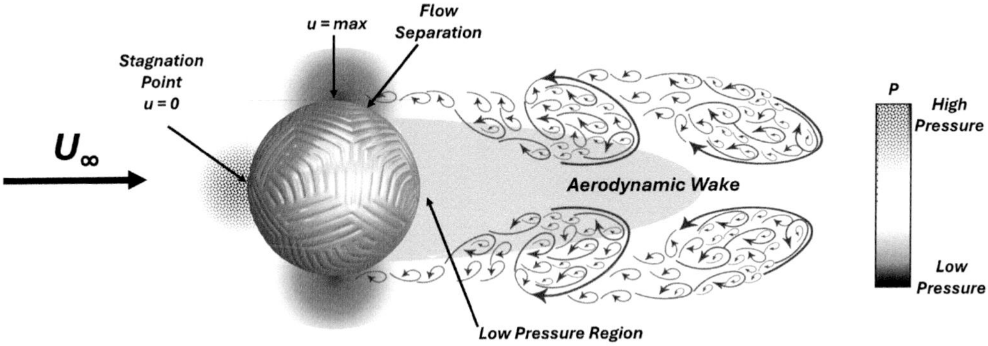

Figure 4.8 A simplified illustration of fluid flow and the pressure field around a ball. At the frontal surface of the ball, flow stagnates, and a high stagnation pressure region forms. Fluid then accelerates around the ball, and at u = max the local pressure falls. After the apex of the ball, surface flow separates from the ball and forms an aerodynamic wake. The swirling flow within the wake causes a reduction in local pressure.

In the preceding sections we have seen that the pressure will vary around a moving object. For example, let's continue to consider the movement of air around a ball in flight. On the front of the ball, pressure increases due to flow stagnation, and we can determine the value of this stagnation using a modified form of Bernoulli's, Equation 4.9. As air flows around the ball, it accelerates to a maximum velocity at the ball's apex, with a corresponding reduction in pressure. The eventual separation of the air flow from the surface of the ball will form an aerodynamic wake. The swirling air within this region, and associated energy dissipation, will cause the pressure to decrease behind the ball. All these pressure forces acting around the ball work together to create pressure drag. Whereas viscous drag acts tangential to the surface of the ball, pressure drag acts normal to it.

The combination of *viscous drag force* and *pressure drag force* acting upon the ball is called the *total drag force*, F_D, the summation of all forces resisting motion due to the fluid:

$$F_D = \frac{1}{2}\rho u^2 A C_D \qquad \text{Equation 4.11}$$

where A is the frontal cross-sectional area of the ball or object, and C_D is a dimensionless number known as the *drag coefficient*. It is more common to write Equation 4.11 in terms of C_D as:

$$C_D = \frac{F_D}{\frac{1}{2}\rho u^2 A} \qquad \text{Equation 4.12}$$

The drag coefficient C_D associated with an object is a function of its Reynolds number but also changes due to the shape, roughness, or texture of the object.

4.5.1 *The influence of Reynolds number on drag*

We can calculate the Reynolds number *Re* using Equation 4.10 to provide an indication of whether the flow conditions for separation within it are classed as laminar or turbulent.

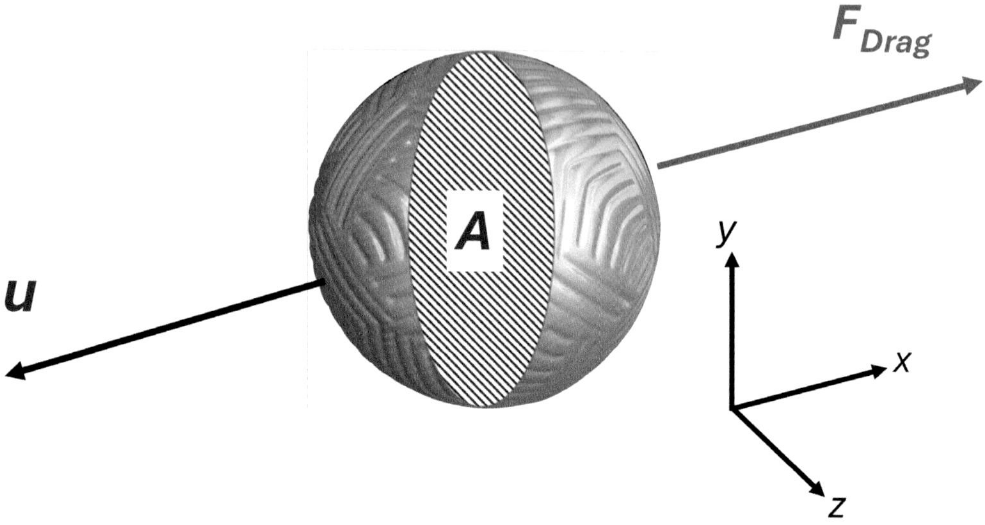

Figure 4.9 Illustration of drag force acting on a ball in opposition to the direction of the balls motion. The shaded region shows the cross-sectional area A.

We will continue to consider the example of flow around a ball. Figure 4.10 shows that the drag coefficient of the ball is dependent on the Reynolds number, and that distinct flow regions exist; subcritical, critical, and supercritical.

When the flow conditions result in a laminar boundary layer separation, flow will detach from the ball's surface before the apex is reached. This point is typically around 80° as measured from the stagnation point. This early separation of the flow results in a large wake forming behind the ball. The combination of frontal stagnation pressure and this large wake result in a high coefficient of drag. For a ball the drag coefficient is typically about 0.5. The range of Reynolds number during which conditions cause a "laminar" boundary separation is termed the "subcritical flow region". If the flow was to accelerate beyond the subcritical Reynolds number range, it would first move into a "critical region" and then beyond into a "supercritical region". Within supercritical flow the boundary layer separations are turbulent. The point at which supercritical flow separates from the surface shifts downstream to around 110° from the stagnation point. This delayed flow separation results in the formation of a smaller wake. Accordingly, the measured drag coefficient reduces to a value that can vary anywhere between about 0.2 to 0.3 depending on the ball's physical characteristics.

Within the "critical flow" region between subcritical and supercritical flow, the boundary layer separation transitions, from laminar to turbulent characteristics. This transition occurs over a relatively narrow Reynolds number range. For a smooth sphere this transition onset occurs around a Reynolds number of about 3.3×10^5. The flow behaviours in this zone are varied, with for example the occurrence of phenomena such as flow separation bubbles where the boundary layer detaches, reattaches, before subsequently detaching again. The Reynolds number range at which the critical transition of an object exists depends on the flow velocity and physical fluid properties and the object's shape and surface characteristics.

4.5.2 The influence of shape on drag

The shape of an object, and its angle of alignment to a flow, will influence its drag coefficient and the Reynolds number when transition occurs. We have discussed how drag coefficient is comprised of viscous drag and form drag components. The contribution that these individual elements make to the total drag is also determined by shape.

It is common to classify the shape of an object as either being streamlined or bluff. Streamlined bodies are associated with low drag coefficients and small flow separations. Examples of such forms include aerofoils. The form drag contribution of a streamline body is low; however, the viscous component is typically high. This is due to flow tending to remain attached, hence a small wake, to the surface of the body. The form drag contribution, however, only remains low if the streamlined body remains aligned to the flow. If the angle of the body increases to the flow, the form drag will also increase.

Bluff bodies, in contrast, have higher total drag coefficients due to large flow separations associated with their commonly blunt, wide forms. The contribution of form drag is therefore high, whilst the viscous drag is typically low due to the low amount of flow remaining attached to the body. Figure 4.11 provides an example of some geometric bodies and typically associated drag coefficients. Hoerner [120] discusses the influence of shape upon drag coefficient for a wide range of classifications of shape.

In sport many objects are classified as having a bluff geometric form, e.g., balls (spheroids), rugby and American footballs (prolate-spheroids), golf drivers (flat faced rounded forms), and shuttlecocks (conical). Human anatomy, although more complex, is also a bluff body that can be deconstructed and simplified to a collection of individual cylindrical, spherical, and prolate spheroid forms. A bicycle is also largely a collection of bluff bodies, although some components of the frame may be streamlined to reduce drag. Understanding aerodynamics that involve humans such as cyclists is complex. The individual bluff geometries that make up the whole interact and therefore can influence localised component forces.

The energy efficiency of bluff bodies with their large flow separations and high drag coefficients is low; therefore in sport effort is often made to mitigate the influence of shape. Cyclists, as discussed in subsection 4.1, may adopt aerodynamically optimised streamlined equipment or postures to either reduce the frontal area of the bluff form or smooth the passage of flow around the shape. Various aerodynamically designed cycling helmets are evident within the sport, with forms that are optimised to both the event and the individual. This approach to customisation is seen across many sports where athletes adopted aerodynamically optimised equipment to increase their speed or performance. Examples include wheelchair racing and downhill skateboarding, and winter sports such as speed skating, skiing, and sled events.

It is not always possible, however, to change the shape of a bluff body. For example, balls need to be round for obvious reasons! However, aerodynamic drag can also be controlled and influenced by the roughness or texture of an object.

4.5.3 The influence of roughness or texture on drag

Sports balls are differentiated visually by their size and surface texture. Golf balls are known for their dimpled surface, which increases their flight distance, tennis balls for

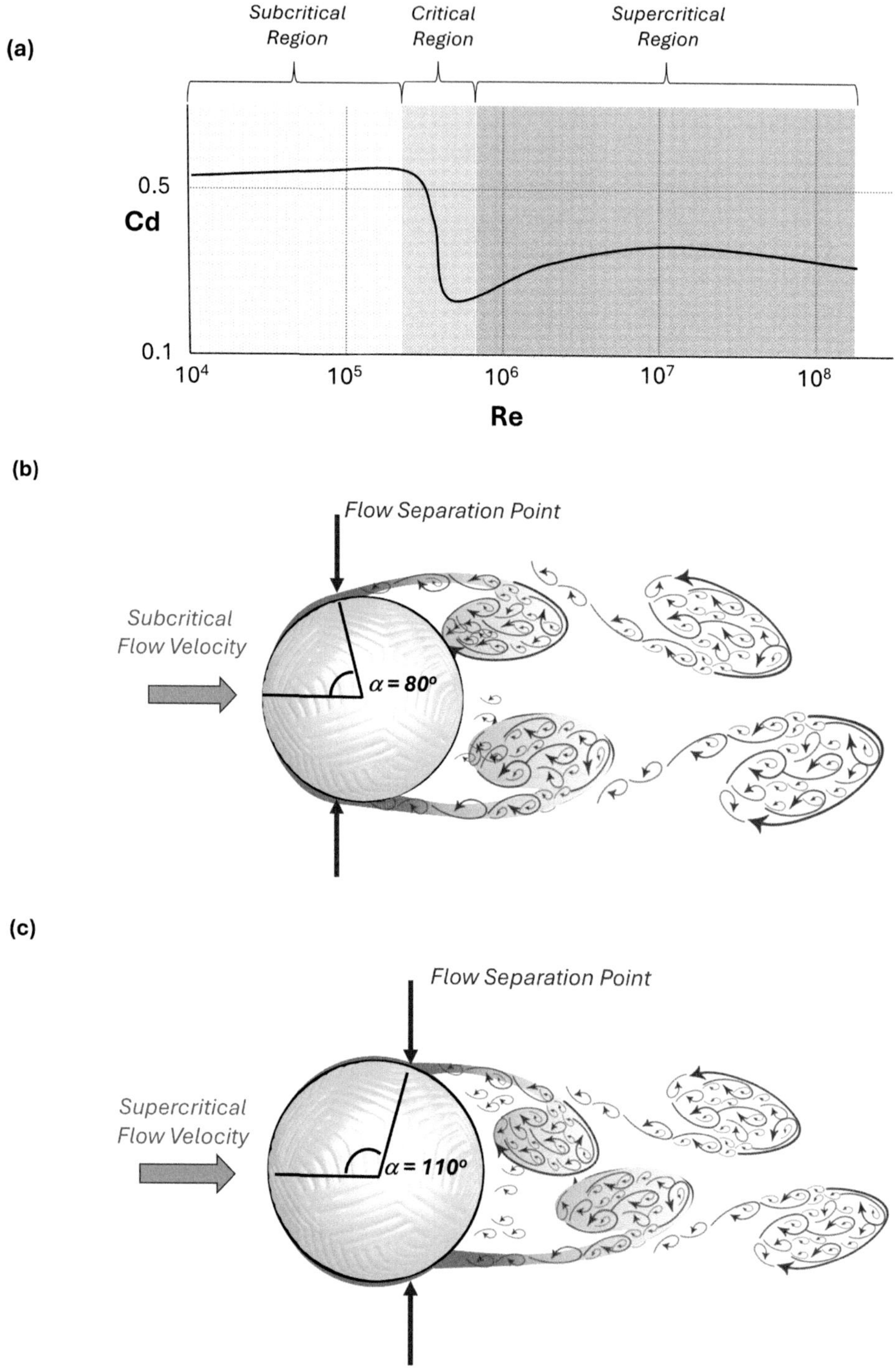

Figure 4.10 (a) Variation of drag coefficient with Reynolds number for a sphere, with three distinct flow regimes, subcritical flow, critical flow, and supercritical flow, (b) illustration of subcritical flow around a sphere, characterised by an early flow separation and large aerodynamic wake, and (c) illustration of supercritical flow around a sphere, characterised by a delayed flow separation and narrower aerodynamic wake.

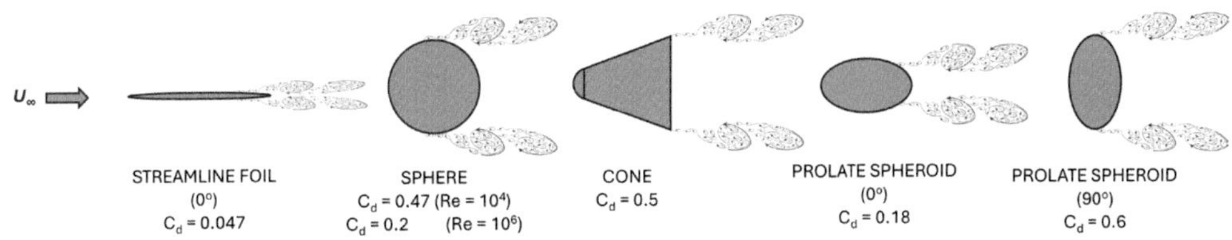

Figure 4.11 Diagram to show the variation of drag coefficient depending on the shape of an object.

their furry nap and high drag coefficient, the patterning of seams on an association football that affect its swerve from a free kick, and the two halves of a cricket ball, separated by a stitched seam that if polished on one side and scuffed on the other will result in a bowling delivery with swing. All these behaviours are attributable to the ball's surface texture influencing the Reynolds number at which flow transition occurs. This translates to the velocity range over which the ball has a subcritical and supercritical drag coefficient. Additionally, as sports balls do not travel at a constant speed, their surface texture can result in interesting behaviours.

Through considered design it is possible to manipulate the Reynolds number at which flow transition will occur around an object by altering surface roughness, texture, and seams. The classical experimental works of Achenbach [121, 122] showed how the onset of flow transition could be manipulated on a sphere by adding glass beads to the surface, creating textures in varying scale. The rougher the surface of the sphere, the earlier the critical flow region was observed to occur, reducing the pressure drag. However, adding surface roughness will increase the skin friction drag that the object experiences.

An example of controlling the flow separation Reynolds number is shown in Figure 4.13. Considering a smooth sphere again, we know that at a subcritical flow velocity its drag coefficient is about 0.5, flow transition occurs at a Reynolds number of around 3.3×10^5, and subsequently its drag coefficient can fall as low as 0.2. By adding surface roughness or a seam, it is possible to shift the transition curve into what could be considered the subcritical Reynolds number range for the sphere. This is achieved by using the seam or texture to increase the energy of the localised flow within the boundary layer, delaying the separation of flow around the sphere and reducing the drag coefficient. It is this behaviour that a highly textured ball, such as a golf ball, utilises to increase its flight distance.

Early golf balls were smooth, but over time surface dimples became an intentional design feature. This was because golfers realised that their older balls, with a worn and roughened surface, flew further. For many sports balls, features that influence surface texture, such as the stitching and seams of panels, were traditionally just manufacturing artefacts. When designing some sports balls nowadays, such as those used for association football, careful consideration is given to the design of the surface texture and associated features.

The textures or seams, used to manipulate transition, need to be carefully designed, as the behaviour of the flow is sensitive to the scale, shape, patterning, and spacing of these applied features. If they are too subtle, they will have little to no effect, and if they are too large, they could have the opposite of the desired effect and cause an early separation of flow.

Figure 4.12 Photographs show (a) golf ball with dimples, (b) tennis ball with a single continuous seam and a hairy surface referred to as nap, (c) football with a moulded surface consisting of seams, and panel texture, and (d) cricket ball with a polished surface and raised, stitched seams.

A classic example of a designed surface texture on a ball having an undesirable effect was the movement seen in the 2010 FIFA World Cup football. In the years preceding the release of this ball, the design of footballs had been rapidly developing. This was due to the introduction of thermal bonding and moulding techniques used in their construction. The classic football design had 32 hexagonal panels with stitched seams. Thermal technology allowed the development of panels of any shape desired, which could also improve the sphericity of the ball. Balls with fewer panels were being developed, with

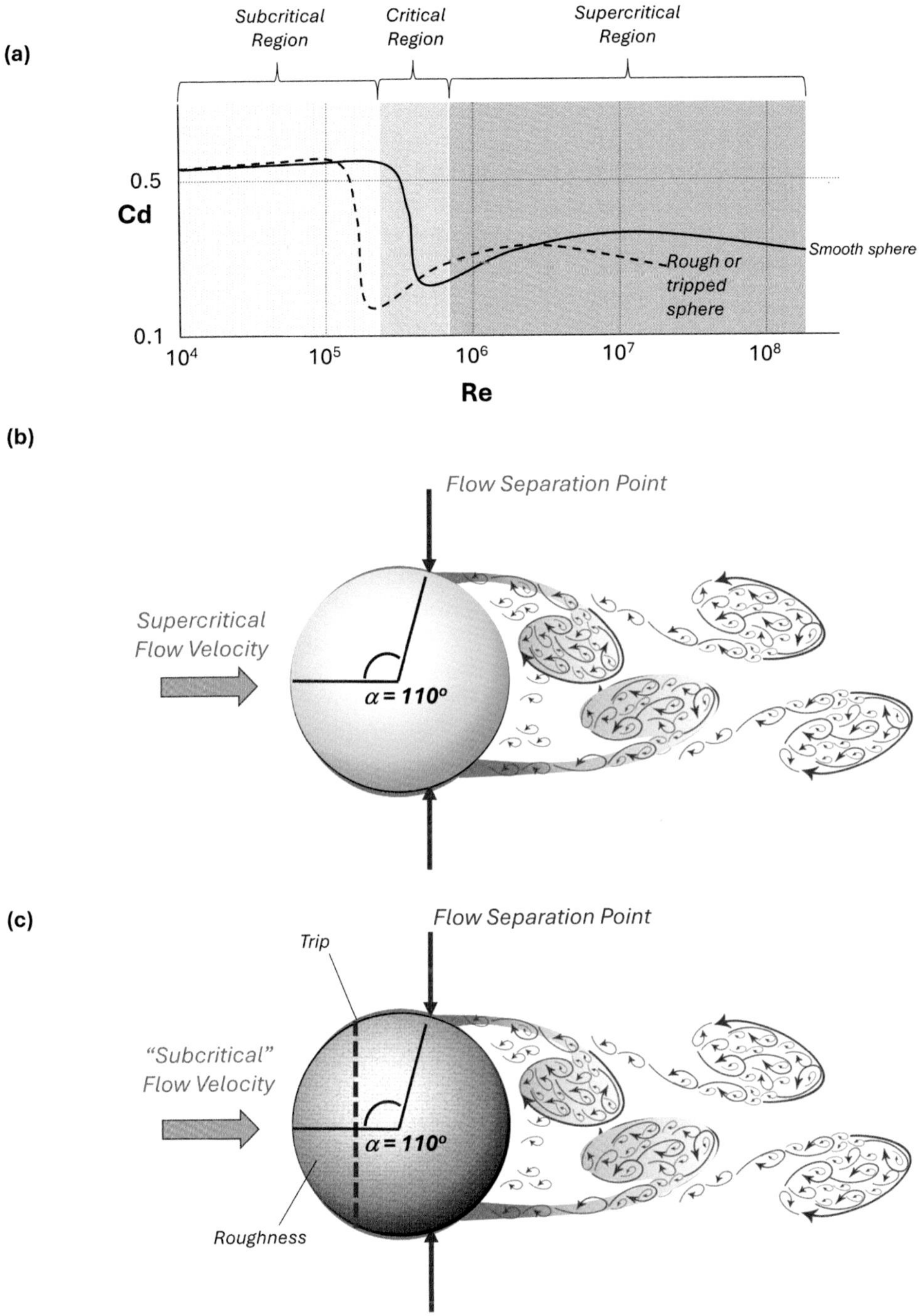

Figure 4.13 Diagram showing (a) variation of drag coefficient with Reynolds number for a sphere, showing the influence of surface roughness on a flow, (b) illustration of supercritical flow around a sphere, characterised by a delayed flow separation and aerodynamic wake of reduced width, and (c) illustration of subcritical flow around a roughened sphere; the surface roughness causes the subcritical boundary layer to transition to a turbulent state, again delaying flow separation and reducing the width of the aerodynamic wake.

the consequence that their surface was becoming smoother. This was due to the reduced overall seam length and depth. Stitched seams were deep, as the stitching pulled the panel edges down and together, whereas thermally bonded seams could almost be hidden if desired.

The 2010 World Cup football had eight panels, and in play exhibited a higher critical Reynolds number than traditional footballs [123]. This meant that it remained within the subcritical flow region for a wider range of Reynolds number than traditional balls and would become erratic in flight at certain kicked velocities. This was noticeable during free kicks where the ball would "float", altering direction mid-flight. This was like a float volleyball serve where the ball shifts unpredictably at lower speeds, in comparison to high-velocity spike shots. This behaviour is not only due to the drag but also the lift and side force experienced by the ball, as discussed in the next subsection (4.5.4).

The 2010 World Cup football became notorious for its aerodynamic behaviours and was criticised by footballers. The ball did have a moulded texture, and "aero grooves" applied to the eight panels, but these were subtle, and insufficient to overcome the small thermally bonded seams. For the following 2014 World Cup football, the manufacturer again used a moulded ball with fewer panels. However, this ball had deeper, longer seams, and a more pronounced texture, to prevent the erratic behaviours of its predecessor. Despite the influence aerodynamics can have on the play of a ball, there is not a specific test for aerodynamic performance within the FIFA quality programme for footballs. The use of surface textures, features, and seams to control flow behaviours goes beyond sports balls and is now widely visible on items ranging from golf drivers, to aerodynamically optimised clothing, helmets, and equipment. These textures have been seen to take many forms in sport, ranging from dimples, pimples, chevrons, trip lines, and assorted vortex generators. They are also becoming increasingly complicated in form, as manufacturing techniques and manufacturers become more adventurous in their application. In recent years, sculpted, grooved textures have been introduced by Nike (Beaverton, USA) on their association footballs, and raised hexagonal honeycomb textures have appeared on Molten volleyballs and lacrosse balls.

4.5.4 *Lift and side forces*

As discussed in preceding subsections, pressure force acting on an object does so in a direction that is normal to its surface. Pressure acting on surfaces normal to the flow direction will create drag. However, as flow moves around an object and the surfaces become parallel to the flow direction, this will result in normal surface forces acting perpendicular to the flow direction. Therefore, additional to drag force an object will experience components of lift force (F_L) and side force (F_S) (Figure 4.14) and are determined as:

$$F_L = \frac{1}{2}\rho u^2 A C_L \qquad \text{Equation 4.13}$$

$$F_{Side} = \frac{1}{2}\rho u^2 A C_{Side} \qquad \text{Equation 4.14}$$

The equations used to describe these forces are similar to that for drag, with the exception that the dimensionless coefficients are now the coefficient of lift (C_L) and the coefficient of side force (C_S). We can rearrange the equations for these coefficients as:

$$C_L = \frac{F_L}{\frac{1}{2}\rho u^2 A} \qquad \text{Equation 4.15}$$

$$C_{Side} = \frac{F_{Side}}{\frac{1}{2}\rho u^2 A} \qquad \text{Equation 4.16}$$

The object will also be subject to the effects of gravity. When lift acts to oppose the direction of gravity, it is termed "lift force". When lift acts aligned with gravity, it is commonly referred to as "downforce".

All bluff bodies will exhibit oscillations in their measured lift and side forces. This is attributable to the boundary layer separating from the body and the associated shed

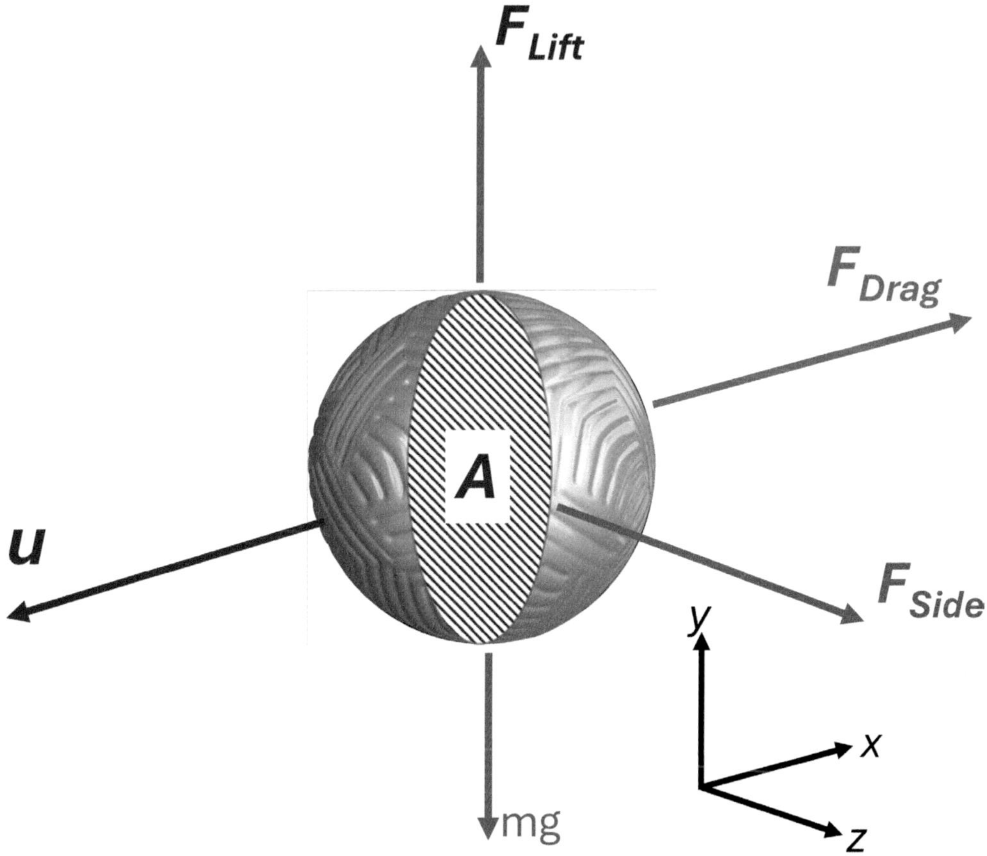

Figure 4.14 Illustration of lift and side forces on a ball that act in a perpendicular orientation to the direction of travel. The shaded region shows the cross-sectional area A.

vortices that make up the wake behind the object. If the body is a ball or cylindrical in shape, the net value of these forces will be negligible, unless the surface texture or seams are asymmetric, i.e., the object is smooth on one side and rough on the other or has a seam such as a cricket ball or baseball.

Figure 4.15 shows a ball with a raised seam. When the seam of the ball is aligned either perpendicular or parallel to the flow, the separation of the wake will be symmetrical with respect to time. Real flow around an object is always time-dependent and the separations will fluctuate both spatially and temporally. However, if the flow is averaged over a sufficient time period, it will average to a symmetrically aligned position. In this scenario any lift or side force generated by the ball will be balanced, equal and opposing. However, if the seam is rotated so that it is inclined to the oncoming flow, the wake separation will become asymmetric. The side of the ball that has a delayed flow separation will have a lower pressure than the side with early separation. This difference in pressure across the ball will cause a net force that acts towards the side with delayed flow separation. Such side forces cause balls in flight to swing.

It is possible to obtain the same swing effect by creating a rough surface on one side of a ball, and a highly polished surface on the other. This will create an asymmetric flow separation around the ball and cause it to swing in flight. Cricketers will try to take advantage of this phenomenon by polishing one side of a ball before delivery. However, in 2018 cricket was rocked by a scandal in an international match between Australia and

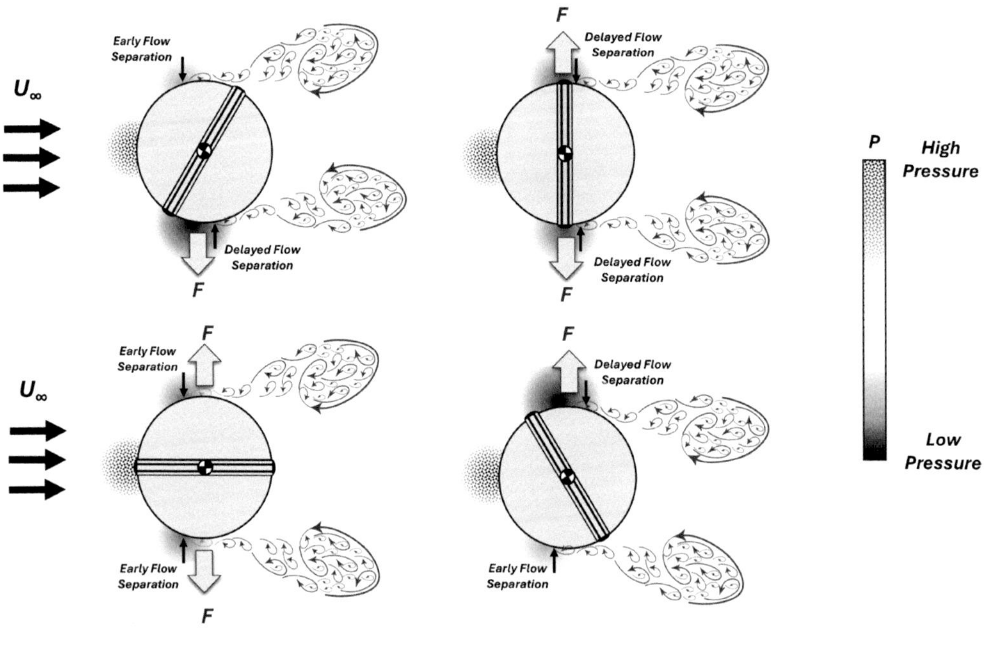

Figure 4.15 Illustration of the influence seam orientation can have on flow separation around a ball and the subsequent orientation of the lift or side force *F*. When the seam here is parallel or perpendicular to the flow, the lift force is balanced on each side of the ball (over an averaged time period). However, if the seam is inclined to the flow, then the separation becomes asymmetric around the ball, creating a net force acting in a single orientation only.

South Africa. The Australian player Cameron Bancroft was caught tampering with the surface of the ball, using sandpaper to artificially condition its surface. This behaviour was ruled as against the spirit of the sport and led to suspension points and fines for both Bancroft and his team captain Steve Smith.

This influence of asymmetric lift or side force is most notable on balls that have minimal or no spin in flight. Baseball pitchers will use it to their advantage to deliver knuckleballs that are unpredictable in trajectory [86, 124]. It is also this behaviour that completes the explanation of the erratic flight of some association footballs, or a float serve volleyball. Asymmetric flow separation, however, is not only evident in balls with asymmetric patterns or textures but also in those with spin.

4.5.5 The influence of spin

Watching baseball or association football is a great way to observe the influence that aerodynamics has on the flight trajectory of a ball. We have already discussed how a ball with minimal or no spin will knuckle in flight. However, if a baseball pitcher or a foot-baller taking a free kick delivers a ball with a high spin, this will cause its trajectory to curve or bend. A classic example is the Brazilian footballer Roberto Carlos's banana free kick against France in 2007. Having positioned the ball 30 yards from goal, Carlos took a 20-yard run up and struck the ball with such power that he imparted a high rate of spin. Although the ball initially appeared to be directed away from goal and thus off-target, the high spin caused it to first curve around the French defensive wall and then continue to curve back in on target and into the net beyond the French goalkeeper. The influence of ball spin is also evident in other sports, e.g., in tennis where high topspin will cause a ball to rapidly bend and dip over a net, or in golf where back spin can extend the flight duration of the ball. A thorough discussion on the wide-ranging behaviours of different sports' projectiles can be found in Goff [125].

The physics that cause the bending motion of spinning balls again is attributable to an asymmetric wake. As the ball spins, an asymmetric flow separation is caused by its rotation. This leads to an early separation on the side of the ball rotating against the flow, and a delayed separation on the side rotating with the flow, as shown in Figure 4.16. This asymmetric separation creates a lift or side force called the Magnus force, which causes the bending motion of the ball.

The effect of spin and Magnus force can be coupled into either Equation 4.13 or Equation 4.14. For example, to couple this into the lift force equation, we first consider a new dimensionless lift coefficient C_l'.

$$C_L' = \frac{C_L}{SP} \qquad \text{Equation 4.17}$$

where spin rate SP is described by:

$$SP = \frac{r\omega}{u} \qquad \text{Equation 4.18}$$

where r is the radius of the ball, w its tangential speed, and again u is the flight velocity. This can be coupled into Equation 4.13 to give:

$$F_L = \frac{1}{2}\rho C_L' r\omega u A \qquad \text{Equation 4.19}$$

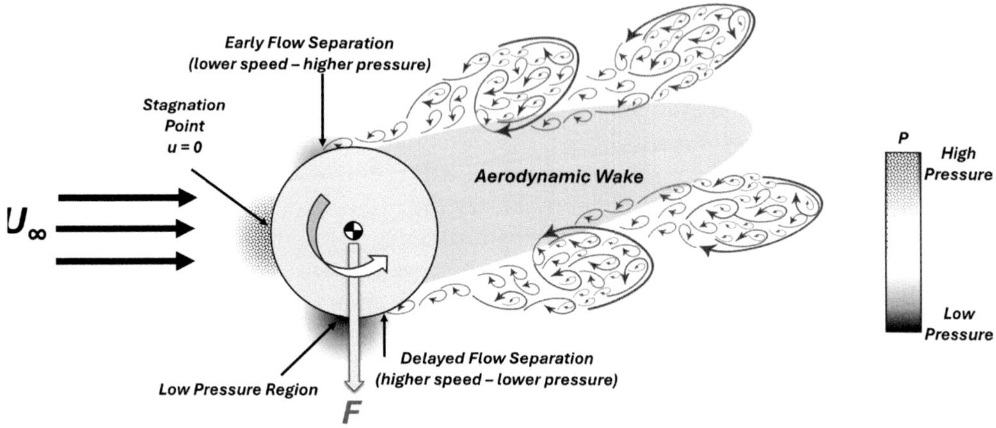

Figure 4.16 Illustration of the influence of top spin on the flow separation around a ball and the subsequent orientation of the lift force *F*.

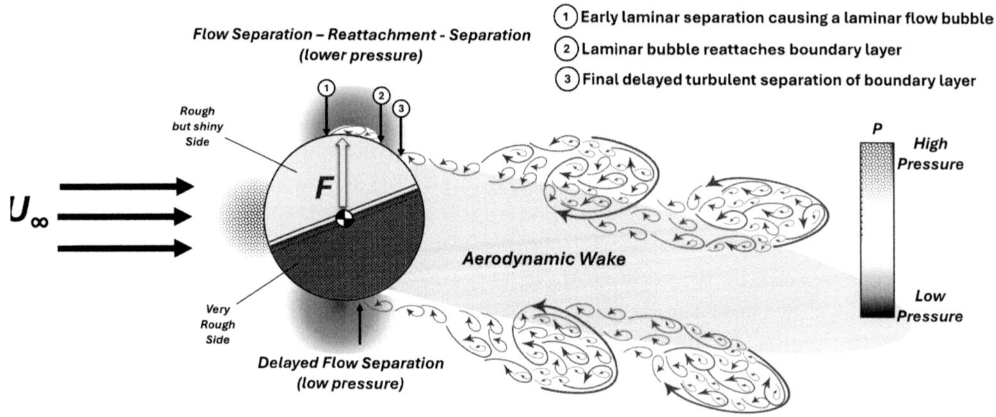

Figure 4.17 Illustration of reverse Magnus, which can be observed in cricket balls travelling faster than 40 m/s and that have a heavily worn surface. The combination of speed and wear causes the formation of a laminar separation bubble on one side of the ball, leading to the separation and subsequent reattachment of flow. This flow eventually separates a second time from the surface of the ball, but this is delayed in comparison to the purely turbulent separation on the opposing side. As a result, the Magnus force reverses and the ball swings in the opposing direction to what may be expected.

There is another scenario, however, where the swing of a spinning ball under the correct conditions will reverse, a condition referred to as reverse Magnus (see Figure 4.17 [126, 127]). This behaviour is evident in cricket. It is difficult to achieve unless the bowler can deliver a ball that is sufficiently worn or old at high speed (>85 mph or ~40 m/s). The behaviour occurs because flow separation is ultimately delayed on both sides of the ball. However, an interesting flow phenomenon is visible, the formation of a laminar separation bubble that occurs in critical flow. Although the boundary layer separates away in a laminar manner, it rapidly transitions to a turbulent state, reattaching to the surface of the ball. It then remains attached up to ~130° where it finally separates. Flow on the opposing side of the ball without a separation bubble detaches earlier at around 110°. Therefore, a pressure difference exists again across the ball, but this time the wake and the direction of the acting Magnus force switch. Further explanation of this phenomenon can be found in Scobie et al. [128].

4.6 How do we understand fluid dynamics?

We have discussed the physics of fluid dynamics that affect the behaviour of sporting goods, and how these behaviours may be utilised and harnessed to a player's advantage, e.g. placing swing on a cricket ball, bending an association football, and knuckling a baseball. We have also seen how these forces can hinder sporting performance. However, through careful design drag forces can be reduced or controlled, which is important in the pursuit of medals or sporting success. To design aerodynamically optimised equipment, we need to build a thorough understanding of aerodynamic theory and the flow physics in our specific problem.

There are three approaches to understanding the fluid dynamics (and mechanics in general) of sporting goods. These are (1) Theoretical, (2) Experimental, and (3) Computational. No single method can provide, or should be relied upon to build, a complete understanding of the fluid dynamic behaviour of sporting goods. They are instead complementary to each other, and all are required to build a comprehensive understanding. All three approaches have different strengths and weaknesses. It is therefore important to consider which approach is most suitable for analysing a problem, considering capabilities and complexity.

So where do you begin? It is always best to start with the use of theory, and there are many texts, such as Hoerner [120] and Anderson [129], that cover theoretical aerodynamics in detail. Use simple analytical equations to start to understand a problem. We provided an example earlier of how Bernoulli's equation can give an indication of stagnation forces acting on an object. We also discussed the limitations of this simple equation. What starts as a simple analytical equation can quickly become cumbersome when applied to the study of flows in multiple directions. However, theory and analytical equations can inform experimental and computational studies.

Experimental techniques are viewed as the "gold standard" for understanding fluid dynamics, and this reputation is probably still deserved. The use of controlled laboratory experimental fluid studies to explore aerodynamic and hydrodynamic behaviours through use of wind tunnels and water flumes can provide rich data sets. This can include both direct force measurements and flow visualisations. However, it can be difficult to extract details of individual component contributions from the whole. Flow visualisation techniques can provide details of separations and boundary layers. However, they can be complex and time-consuming to set up. A further limitation arises from the artificial constraining boundaries that a flow tunnel or flume creates and that must be accounted for in measurements; and that an object needs to be physically attached to a piece of equipment to obtain force measurement. The alternative to the laboratory is to conduct experiments in the field, but this again brings challenges to understand both the flow physics and resulting forces. Technological developments in sensors and flow visualisation are enhancing research possibilities.

The third approach is the use of simulation or Computational Fluid Dynamics (CFD). It is an established and embedded tool within commercial sporting goods design, elite sport technical programmes, and academic sports research. This is for the simple reason of the powerful insight and associated benefits to product development and understanding CFD delivers. This has become possible due to advances in computing capabilities and the increased accessibility of affordable high-performance computing. However, challenges still exist, including the accurate prediction of transitional flow behaviours, which as we have seen throughout this chapter are key to all aerodynamic behaviours. CFD, however, is still valuable due to the rich and detailed flow physics that can be uncovered if it is applied in a careful and considered manner. It is essential that a user of the technique understands its capabilities and limitations, as discussed by Hart in Chapters 3 and 4 of ref. [10].

In conclusion, it is important that no one technique should be relied upon to provide a complete understanding of fluid dynamic behaviour. The individual approaches should not be used in isolation but be layered to build a comprehensive and full picture.

4.7 Chapter summary

This chapter has provided a broad overview of fluid dynamics in sport. It began with an overview of why fluid dynamics is an important consideration for sports engineers and introduced principal fluid dynamics concepts. The application to sporting scenarios was discussed throughout, with emphasis on how we can control aerodynamic behaviours through considered design of surface textures, patterns, and seams. Readers should now have a basic understanding of the role and importance of fluid dynamics to sports engineers. This basic understanding should be sufficient to see how fluid dynamics relates to the complementary topics covered in this introductory textbook on sports engineering. However, there is still more to be learnt about fluid dynamics beyond the basics covered here. Students are encouraged to continue reading and learning about fluid dynamics and how it applies to sport.

4.8 Suggested further reading

The following suggestions for further reading go deeper into the topics introduced in this chapter:

- *Fluid Mechanics* [116]
- *Gold Medal Physics: The Science of Sports* [108]
- *Routledge Handbook of Sports Technology and Engineering*, 1st and 2nd editions [10, 86]
- *Gliding for Gold: The Physics of Winter Sports* [109]
- *Mechanics of Fluids* [117]
- *Physics of Baseball and Softball* [92]

5 Fundamentals of sports materials[1]

Materials are key to sports engineering and sporting goods design. Advances in materials have improved sporting goods and facilities. Such advances allow athletes to perform better and make participation in sport more accessible. For example, uptake of fibre-matrix composites has improved various sporting equipment, such as skis and snowboards, tennis rackets [30], bike frames, and more recently running shoes. Developments in materials have also led to better protective equipment and devices, which can enhance safety, particularly in contact and action sports.

This chapter introduces the important topic of sports materials. It covers the fundamentals of sports materials. Basic materials topics are introduced and discussed in the context of sport and sporting goods.

5.1 Density

Density is an important consideration in the material selection for sporting goods. As introduced in Chapter 4, the density of a material is the ratio of its mass to volume. This means if two material samples were the same size, but one was twice as heavy, it would be twice as dense. Low-density materials are often desirable for sporting goods – think lightweight running shoes or racing bikes. In many cases, materials are selected based on their ratio of density to a basic mechanical property, such as stiffness or strength (more to follow). Considering materials commonly used in sporting goods, metals tend to be denser than polymers, which in turn tend to be denser than foams (i.e., density of metals > polymers > foams).

Imagine a 10 × 1 × 1 cm polymer sample with a mass of 10 g. Such a sample would have a volume of 10 cm^3, and a density of 1 g/cm^3.

5.2 Stress

Stress (σ) is:

$$\sigma = \frac{F}{A} \qquad \text{Equation 5.1}$$

where A is the cross-sectional area of the material. It is the force per unit area being applied to a material. The SI units of stress are N/m^2, which are often referred to as pascals (Pa). There are different types of stress, depending on how the load is applied and the sample is constrained, including bending, compressive, shear, tensile, and torsional. While the applied load will cause a change in the cross-sectional area of the material,

DOI: 10.4324/9781003411086-5

engineers typically use the original dimensions when calculating stress during material characterisation tests. This is called "engineering stress".

Stress is important for defining the behaviour and performance of materials and structures such as sporting goods, as described later in this chapter. Returning to our hammer thrower in Chapter 3, the opposing centripetal and centrifugal forces will tension the wire. The stress in this wire will depend on its cross-sectional area and the tensile force (caused by the centripetal/centrifugal forces).

Our 10 × 1 × 1 cm polymer sample would have a cross-sectional area of 1 cm^2. This means if an axial tensile force of 500 N were applied, the engineering stress would be 5 MPa.

5.3 Strain

Strain (ε) is:

$$\varepsilon = \frac{\Delta L}{L} \qquad \text{Equation 5.2}$$

where L is length. It is the proportional change in length of a material under an applied load. As with stress, Equation 5.2 is often applied to the original, rather than instantaneous, length of the sample. It is then referred to as "engineering strain". If we add small increments of strain, calculated using instantaneous values, then we no longer have the same value as in Equation 5.2. Like stress, there are different types of strain, depending on how the load is applied and the sample is constrained. While strain is unitless, it can be expressed as a percentage when multiplied by 100.

Stretch is sometimes used in place of strain for tensile load. It is defined as the ratio of current to original length. Taking the natural logarithm of stretch gives a quantity usually known as true strain. This term has some advantages over engineering strain. Whether we add small increments where we change L or consider a large deformation where we divide by original L, the calculated value of true strain does not change. For this reason, sports engineers working with soft materials that stretch easily, like foam or tissue simulants, will often use true strain.

If our 10 × 1 × 1 cm polymer sample was stretched to a length of 10.1 cm, the engineering strain would be 0.01 (or 1% strain), according to Equation 5.2.

5.4 Elastic modulus

Elastic modulus (E), or modulus of elasticity, is:

$$E = \frac{\sigma}{\varepsilon} \qquad \text{Equation 5.3}$$

It is a key material property that determines performance when stretched or compressed. Elastic modulus is analogous to material stiffness. It can be used to predict how a material or structure will deform under a given load and forms the basis of stress analysis. As our polymer sample had an engineering strain of 0.01 for an engineering stress of 5 MPa, according to Equation 5.3 its elastic modulus would be 0.5 GPa.

Elastic modulus is typically determined from a tensile or compressive test on a material sample. Young's, or tensile, modulus, is specifically the modulus of elasticity in tension.

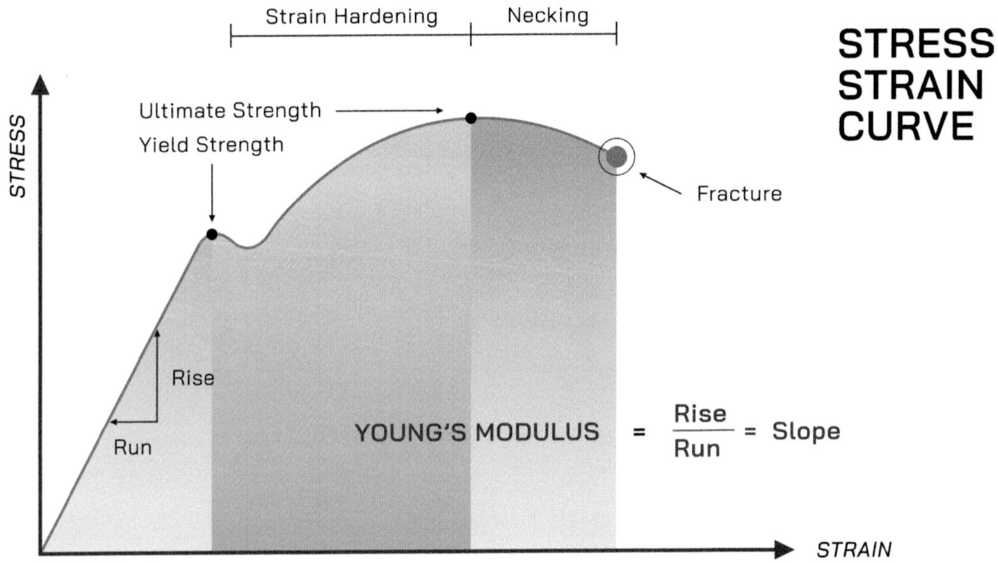

Figure 5.1 A graph showing a typical stress–strain curve for a tensile test on a ductile material. The area under the curve represents the energy absorbed per unit volume.

Figure 5.1 shows a typical stress–strain relationship for a tensile test on a ductile material. Young's modulus can be obtained from the gradient of the initial linear region of the stress–strain curve (i.e., the ratio of stress to strain). One way of obtaining the gradient of this initial linear region is to fit a straight line to it.

5.5 Strength

Young's modulus only applies to the initial linear elastic region of the stress–strain curve. The yield point often marks the transition from the elastic to the plastic region. The yield strength corresponds to the stress at the yield point (i.e., the maximum stress a material sample can take before deforming plastically) (Figure 5.1). If the stress is within the elastic region (i.e., below the yield stress) the material will return to its original size when the load is removed. If the stress is within the plastic region (i.e., above the yield stress), it will have plastic, or permanent, deformation when the load is removed.

Ultimate strength (Figure 5.1) is the force required to break the chemical bonds that hold a material together. Interestingly, these have almost no correlation to the actual strength of the chemical bonds, which (historically) caused confusion for engineers and scientists. Instead, material strength is related to the way in which stress concentrates around small cracks in the material that are present in nearly everything. Fibre-matrix composite materials are a rare exception to this rule. Their fibres are so thin that they can be made with very few cracks, meaning that they can be extremely strong. Interestingly, it was the discovery of the high strength of fibres that eventually helped scientists solve the puzzle of material strength, and why it appeared not to relate to chemical bond strength. So, these fibre-matrix materials are an excellent example of a search for understanding leading to the design of better products.

Brittle materials exhibit minimal plastic deformation, which means they can fracture, break, snap, or crack suddenly and catastrophically if the stress exceeds the yield value.

Figure 5.2 A photo of a surfer with a snapped surfboard.

Fibre-matrix composite materials can be brittle, particularly if they contain defects, meaning they can snap suddenly. Surfboards have a foam core surrounded by a composite skin. Imagine a surfboard snapping during a fall on a large wave (e.g., Figure 5.2).

Considering materials commonly used in sporting goods, metals have a higher Young's modulus than polymers, which in turn typically have higher values than foams (Young's modulus of metals > polymers > foams). Metals also tend to be stronger than polymers, which in turn tend to have higher strength than foams (strength of metals > polymers > foams).

5.6 Specific modulus and specific strength

Specific modulus, or specific stiffness or stiffness-to-weight ratio, is:

$$Specific\,modulus = \frac{E}{\rho} \qquad \text{Equation 5.4}$$

The SI units of specific modulus are Nm/kg (or m^2/s^2). Specific modulus is useful for determining materials that are both light and stiff. This can be important for sporting goods that must be light and resist deformation, such as bike frames and tennis rackets.

Specific strength, or strength-to-weight ratio, is defined as:

$$Specific\ strength = \frac{Strength}{\rho} \qquad \text{Equation 5.5}$$

The SI units of specific strength are the same as for specific modulus (Nm/kg or m^2/s^2). Specific strength is useful for determining materials that are both light and strong. This can be important for sporting goods that must be light and not break easily, such as bike frames. The fibres used in fibre-matrix composite materials, such as glass and carbon fibre, have high specific strength (see Table 5.1). Hence composite materials are often used in sporting goods.

Material selection charts, commonly known as Ashby plots, are a convenient and systematic way to visually compare the properties of different materials. Ashby plots are often used for material selection, including within dedicated software, and are well documented in the literature (e.g., [130]). The two examples in Figure 5.3 plot (a) Young's modulus against density and (b) tensile strength against density for some types of materials used in sporting goods. Note the log scales on the axes of these plots, as the displayed properties vary considerably between materials. These examples help to visualise the concept of specific stiffness and specific strength.

The foams in the bottom left of both plots have a low stiffness, strength, and density. The metals and alloys in the top right of the plots have a high stiffness, strength, and density. Note the white spaces in the top left and bottom right of the plots. Top left is a desirable region, as for plot (a) it corresponds to high stiffness and low density (i.e., high specific stiffness) and for plot (b) high strength and low density (i.e., high specific strength). Materials can be specifically engineered to fall within these regions, for specific sporting applications. Metamaterials are an example of such engineered materials, which are described at the end of this chapter.

5.7 Poisson's ratio

Poisson's ratio (v) is:

$$v = -\frac{\varepsilon_{la}}{\varepsilon_{lo}} \qquad \text{Equation 5.6}$$

where the subscripts *la* and *lo* refer to lateral and longitudinal, respectively (see Figure 5.4). As such, Poisson's ratio refers to the amount a material contracts or expands sideways when stretched or compressed along its length. For 3D isotropic materials (covered later), Poisson's ratio can be between –1 and 0.5. Conventional materials have a positive Poisson's ratio, meaning they get thinner when stretched lengthwise (hence the minus sign in the equation) (see Figure 5.4). Auxetic materials have a negative Poisson's ratio, meaning they get fatter when stretched lengthwise (see Figure 5.4, ref. [23], and the metamaterials section).

Let's return to our 10 × 1 × 1 cm polymer sample. If it contracted laterally by 0.04 mm when stretched longitudinally by 1 mm, its lateral and longitudinal engineering strain would be –0.004 and 0.01, respectively (see Equation 5.2). So according to Equation 5.6, it would have a Poisson's ratio of 0.4.

A Poisson's ratio of 0.5 means the volume of the material would not change when stretched. This is because the volume gain from the increase in length equals the loss from the decrease in cross-sectional area. Such materials are termed incompressible, as the volume cannot be changed by mechanical deformation. While rubbers can have a Poisson's ratio close to 0.5, truly incompressible materials do not exist. All other Poisson's ratios in the range –1 to 0.5 (limits for isotropic materials) cause the material to increase in volume when stretched. A Poisson's ratio of 0 means the

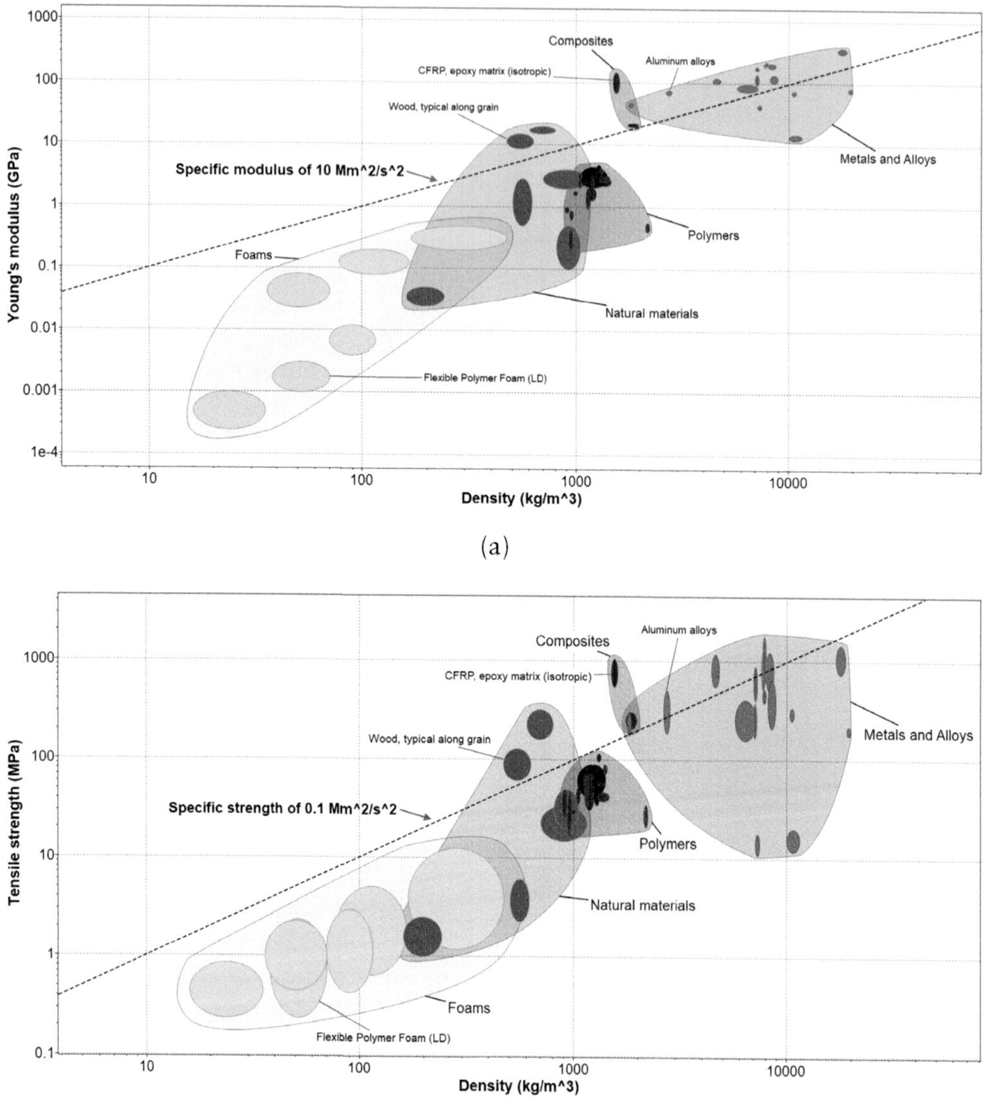

(a)

(b)

Figure 5.3 Ashby plots showing (a) Young's modulus vs. density and (b) tensile strength vs. density. The dashed line in each plot corresponds to a constant (a) specific modulus of 10 Mm^2/s^2 and (b) specific strength of 0.1 Mm^2/s^2. Plots generated by the author in Granta software to illustrate the properties of some of the types of materials used in sporting goods. CFRF is carbon fibre-reinforced plastic and LD is low-density. Images used courtesy of ANSYS, Inc.

cross-sectional area of the material will not change when it is stretched or compressed longitudinally.

Polymers tend to have higher Poisson's ratios than metals (e.g., typically >0.35 vs. ~0.25 to 0.35). Cork has a Poisson's ratio close to zero. The Poisson's ratio of wood varies in different directions (~0.2 to 0.4) (see upcoming section on anisotropic materials).

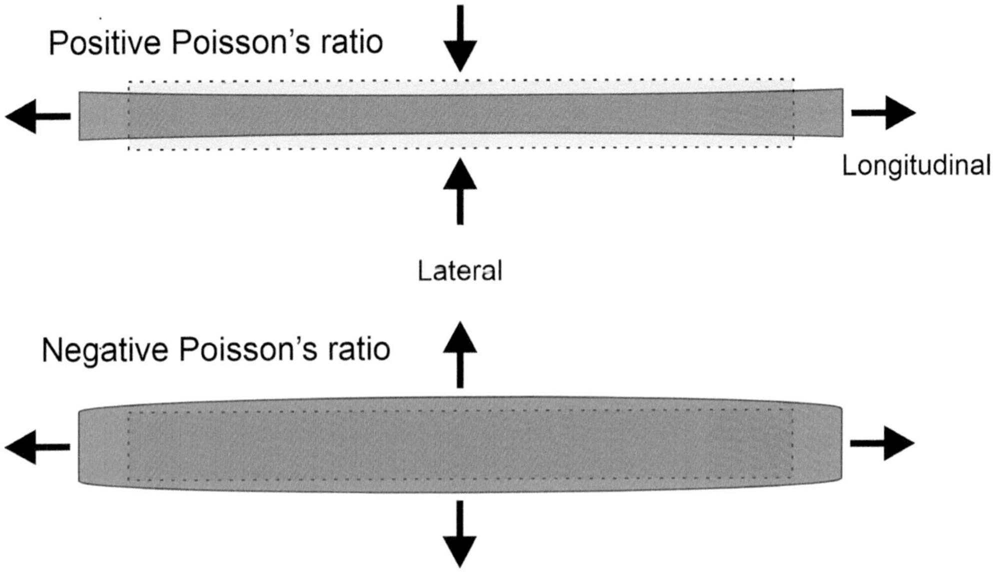

Figure 5.4 Illustration showing how materials with positive (top) and negative (bottom) Poisson's ratio deform laterally when stretched longitudinally.

5.8 Shear modulus

Shear modulus (G) is the ratio of shear stress to shear strain. It is analogous to material shear stiffness, and is related to modulus of elasticity and Poisson's ratio as follows:

$$G = \frac{E}{2(1+v)} \qquad \text{Equation 5.7}$$

According to Equation 5.7, shear modulus will increase with modulus of elasticity and as Poisson's ratio decreases. A characteristic of auxetic materials is that shear modulus will tend towards infinity as Poisson's ratio tends towards –1. This is because all the deformation happens by volumetric deformation.

5.9 Bulk modulus

Bulk modulus (K) is:

$$K = -\frac{PV}{\Delta V} \qquad \text{Equation 5.8}$$

where P is pressure. The minus sign is because an increase in pressure causes a decrease in volume. Bulk modulus dictates the ability of a material to withstand a change in volume under hydrostatic pressure (compression on all sides). It is related to modulus of elasticity and Poisson's ratio as follows:

$$K = \frac{E}{3(1-2v)} \qquad \text{Equation 5.9}$$

According to Equation 5.9, bulk modulus will increase with elastic modulus and Poisson's ratio. As Poisson's ratio tends towards 0.5, bulk modulus tends towards infinity (i.e., an incompressible material). A characteristic of auxetic materials is that bulk modulus becomes very low as Poisson's ratio tends towards −1. This means auxetic materials have low resistance to volume change, e.g., when stretched longitudinally auxetic materials stretch in all three axes and quickly increase their volume (Figure 5.4).

5.10 Hardness

Hardness is defined as the ability of a material to resist localised deformation, which usually means resistance to permanent plastic deformation. Harder materials are typically more resistant to surface damage via indentation, scratching, or abrasion. This means sporting goods made of harder materials will tend to have a more wear-resistant surface. For example, adding a hard surface coating to a bike frame could enhance its scratch resistance. Elsewhere, skis or snowboards with a harder base will be more resistant to scratching and abrasion from contact with objects like protruding rocks or from sliding on boxes or rails during tricks. Hardness can also affect how well a material will "grip" a surface (see friction in Chapter 3). For example, the hardness of rubber on the sole of a climbing shoe will influence how rough features on a rock face (asperities) protrude into it and hence how well it grips (see ref. [104]).

There are various tests for measuring hardness, as described in standards such as "ISO 48-1:2018 Rubber, vulcanized or thermoplastic – Determination of hardness – Part 1: Introduction and guidance". Shore hardness is a common method for measuring a material's resistance to indentation, as measured with a durometer (see "ISO 48-4:2018 – Rubber, vulcanized or thermoplastic – Determination of hardness – Part 4: Indentation hardness by durometer method (Shore hardness)"). There are various Shore hardness scales that can be used for comparing the hardness of different materials. A higher Shore hardness corresponds to a harder material with greater resistance to indentation. Shore A and Shore D hardness scales are the most common, with the latter typically used for harder materials than the former.

Climbing shoe rubber would typically have a hardness ranging from 60~75 Shore A, skateboard and rollerblade wheels would typically have a hardness ranging from 75~98 Shore A, and a golf ball would typically have a hardness of 34~73 Shore D (depending on the measurement location) [86, 104, 131]. Shore hardness, or durometer, values often accompany skateboard wheels to help skateboarders choose those that align to their riding style and preference. As an example of the use of hardness in sports governance, the International Bobsleigh & Skeleton Federation specify that the hardness of each runner must not exceed 385 Brinell [132]. Shore hardness is often provided on material datasheets, including those for additive manufacturing, while elastic modulus may not be. This can make selecting the right material challenging when part stiffness/flexibility is a requirement.

5.11 Isotropic and anisotropic

Isotropic materials exhibit the same mechanical response when loaded in different directions (Figure 5.5). For example, an isotropic material would have the same elastic modulus regardless of the measurement direction. Examples of materials used in sporting goods that tend to be isotropic include most plastics and metals. In contrast, anisotropic materials exhibit a different mechanical response when loaded in different directions, i.e., they are directionally dependent (Figure 5.5). This means properties such as elastic modulus depend on the measurement direction. Orthotropic materials are a subset of anisotropic materials, and their mechanical response to loading changes in three mutually perpendicular directions, i.e., they have three planes of symmetry. Wood is an example of an orthotropic material used in sports equipment (Figure 5.5), such as baseball and cricket bats and skateboard decks. It is stiffest and strongest along the grain fibre direction.

5.12 Homogenous and heterogeneous

A homogenous material has uniform properties throughout (Figure 5.5). For example, the density of a homogenous material will be the same regardless of the measurement location. Inhomogeneous, or heterogenous, materials have non-uniform properties. For example, properties like their density and hardness could vary with the measurement location. As a natural material, wood is a classic example of a heterogeneous material used for sporting goods.

5.13 Typical material properties and sporting applications

Table 5.1 provides the properties of some common materials, alongside examples of where they are used in sporting goods. The properties of such materials can vary with the processing conditions, and hence those provided here are just indicative. Table 5.1 highlights the importance of specific properties. For example, while steel alloys are almost three times stiffer than aluminium alloys, they are also almost three times denser, so their specific modulus is similar. Note the specific modulus of titanium alloy is also similar to that of steel and aluminium alloy, as it has intermediate Young's modulus and density. Fibre-matrix composites feature in Table 5.1 as they are used in various sporting goods, as detailed in the next chapter.

5.14 Material characterisation methods

There are various ways to characterise materials used in sporting goods. There are also various standards for testing materials. These standards define things like the size and shape of test samples, the test speed, the environmental conditions (temperature and humidity), and how to report the results. Examples of such standards include:

- ISO 6892-1:2019 Metallic materials – Tensile testing Part 1: Method of test at room temperature
- ISO 604:2002 Plastics – Determination of compressive properties
- ISO 14125:1998 Fibre-reinforced plastic composites – Determination of flexural properties

There are many advantages of following standard tests, such as easier communication, reporting, and comparison of test results between different studies, laboratories, and material suppliers. For example, material suppliers typically provide key material properties on a corresponding data sheet alongside the code of the standard, e.g., ISO 6892-1:2019, followed to test them. This reduces the detail, i.e., test method descriptions, needed on

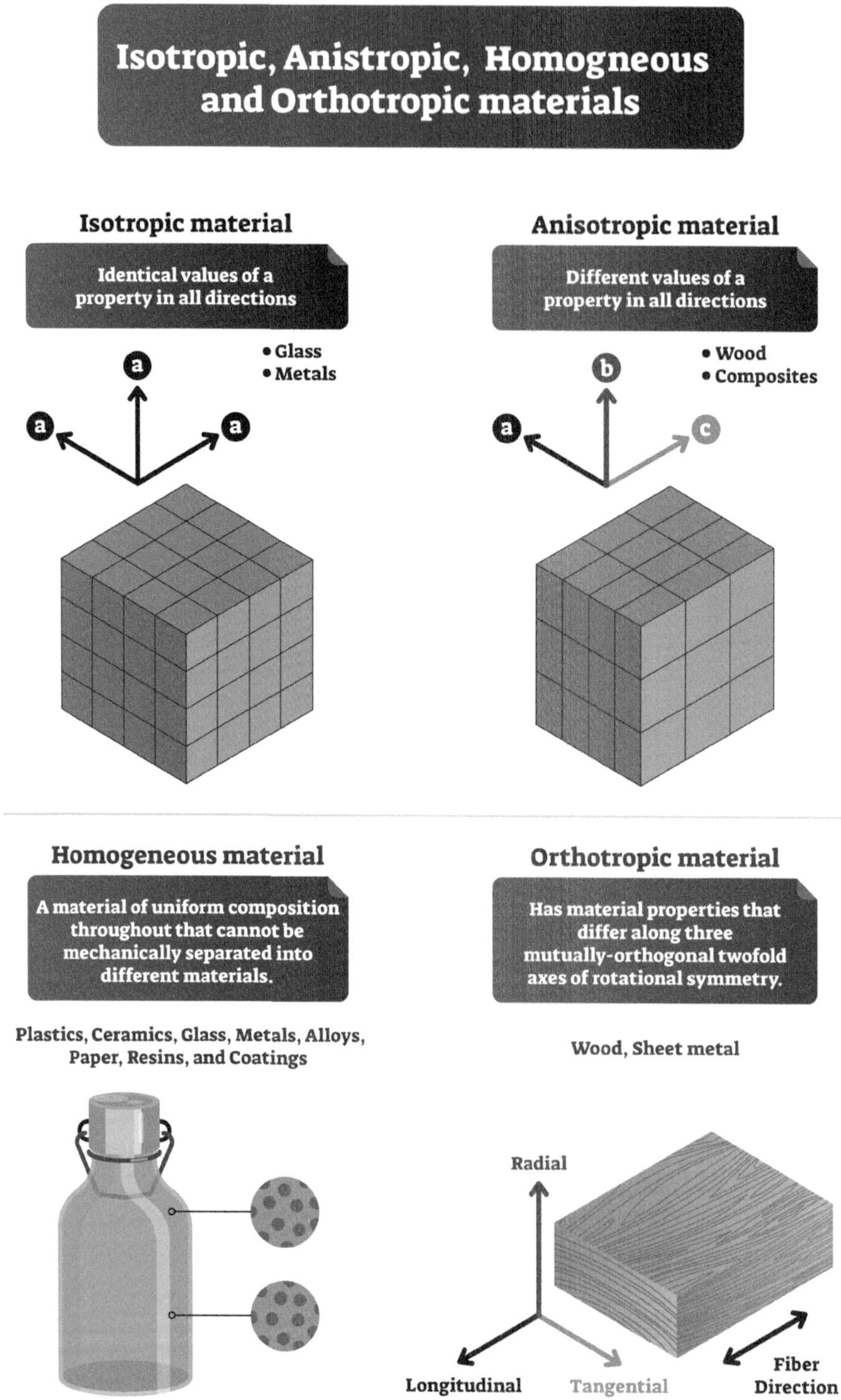

Figure 5.5 Illustration showing isotropic, anisotropic, homogenous, and orthotopic materials.

Table 5.1 Typical properties of some materials commonly used in sporting goods, collated from refs [28, 86, 133–136] and Granta software. These are just indicative, and when specific values are required, they should be obtained from the literature or via testing

Material	*Density (g/cm^3)*	*Young's modulus (GPa)*	*Tensile strength (GPa)*	*Specific modulus (Mm^2/s^2)*	*Specific strength (Mm^2/s^2)*	*Example sporting applications*
Alloys*						
Aluminium	2.7–2.8	69–75	0.2–0.6	25–28	0.07–0.22	Bike frames and components, baseball/softball bats
Steel	7.8–7.9	195–210	0.4–1.8	25–27	0.05–0.23	Bike frames and components, ski/snowboard edges, ice-skate blades
Titanium	4.4–4.7	85–125	0.8–1.4	18–28	0.2–0.3	Golf clubs
Polymers						
Nylon/ Polyamide 6	1.1–1.2	1.4–2.8	0.04–0.08	1.2–2.5	0.03–0.07	Ski and snowboard bindings, ski boots, inline skates, footwear components
Polycarbonate	1.2	2.3–2.4	0.06–0.07	1.9–2.0	0.05–0.06	Helmets, ice hockey face shields
Woods**						
Willow	0.3–0.4	6.6–8.7	0.05–0.06	17–29	0.13–0.20	Cricket bats
Ash	0.5–0.6	9.7–12	0.09–0.10	16–25	0.15–0.21	Baseball bats, oars
Maple	0.5–0.6	11.3–12.6	0.09–0.11	18–23	0.15–0.20	Baseball bats
Fibre-matrix composites**						
Constituent materials						
Epoxy resin/ Polymer matrix	1.1–1.4	1–6	0.04–0.1	0.7–5	0.03–0.09	Bike frames and components, tennis rackets, golf clubs, fishing rods, canoe/kayak paddles, skis and boards, boats, plates in running shoes
E-glass fibre	2.5–2.9	70–115	2.0–3.4	24–46	0.7–1.4	
Carbon fibre	1.7–2.2	215–650	1.9–3.4	98–382	0.9–2.0	
Aramid fibre	1.4–1.5	58–186	2.8–3.0	39–134	1.9–2.1	
Unidirectional fibre composites, ~60:40 fibre-epoxy ratio***						
E-glass-epoxy	2.0–2.1	42–45	1.0	20–23	0.5	
Carbon-epoxy	1.5–1.7	130–360	1.2–2.1	76–240	0.7–1.4	
Aramid-epoxy	1.4	76–77	1.4–1.8	54–55	1.0–1.3	

* An alloy contains at least two substances, at least one of which is a metal. There are various types of alloys, for which the properties differ.

** Young's modulus and strength parallel to the fibres.

*** Properties will vary depending on the actual fibres and epoxy resin, the fibre-resin ratio, and the direction of the fibres.

data sheets and allows someone interested in using the material in sporting goods to quickly gauge the properties. They can also easily compare them to alternatives, including those of other suppliers.

Following the correct standard test also reduces the likelihood of common mistakes. For example, while many tensile testing standards suggest "dog-bone" shaped samples (e.g., like those in Figure 5.6), to better spread clamping forces, those for continuous fibre composites do not [137]. This is because cutting such samples from a sheet of composite material would mean severing the fibres where the sample widens towards each end, resulting in stress being incorrectly transferred from the test device clamps. Standards are covered further in Chapter 8. Some common themes and methods of testing materials are now summarised.

When testing materials, sports engineers commonly use a universal test device (see Figure 5.6a) to load a sample of known dimensions while measuring its displacement. The sample would typically be loaded slowly in tension or compression. Force and displacement measurements from such a test are used to calculate engineering stress and strain from measurements of initial dimensions (using Equation 5.1 and Equation 5.2). Key material parameters, such as Young's modulus (Equation 5.3) and yield and ultimate strength, can be extracted from a stress–strain plot (e.g., Figure 5.1). Tangent modulus can also be obtained as the ratio of stress to strain throughout the test, rather than just the initial linear elastic region as with Young's modulus. Samples for tensile tests are typically "dog-bone" shaped (see Figure 5.6b), while those for compression tests are typically "disk-shaped". Some standards suggest that friction between compression samples and the test platens should be minimised, such as by lubricating them, to prevent the sample from forming a barrel shape ("barrelling").

Flexural, or bend, tests are also commonly used by sports engineers for characterising materials. Common types include two-, three- (Figure 5.7), and four-point bend tests. For these tests, a specific rig can be mounted in a universal test device. Flexural tests can be useful when tensile tests are challenging, e.g., brittle, or "hard-to-grip" materials, or when bending properties are of particular interest, e.g., materials for use in sports equipment that will be subjected to bending, such as skis and snowboards or baseball bats. The test specimen is typically a rectangular cuboid (Figure 5.8). Using its known dimensions, key material properties can be calculated. For example, for rectangular cuboid samples in a three-point bend test, flexural modulus (E_f) is:

$$E_f = \frac{L^3 m}{4bd^3} \qquad \text{Equation 5.10}$$

where L is the support span (distance between the two supports), m is the gradient of the load-deflection plot, and b and d are the width and depth of the sample, respectively. Similarly, flexural stress (σ_f), and hence strength, is defined as:

$$\sigma_f = \frac{3F_r L}{2bd^2} \qquad \text{Equation 5.11}$$

where F_r is the load at rupture/fracture.

During uniaxial tests, i.e., tension or compression, loads and strain are evenly distributed on the cross-section of the sample. Close to compression plates or test device jaws, however, friction and gripping cause non-uniform stresses. In contrast, during flexural tests, stress and strain within the sample vary with the distance from its neutral

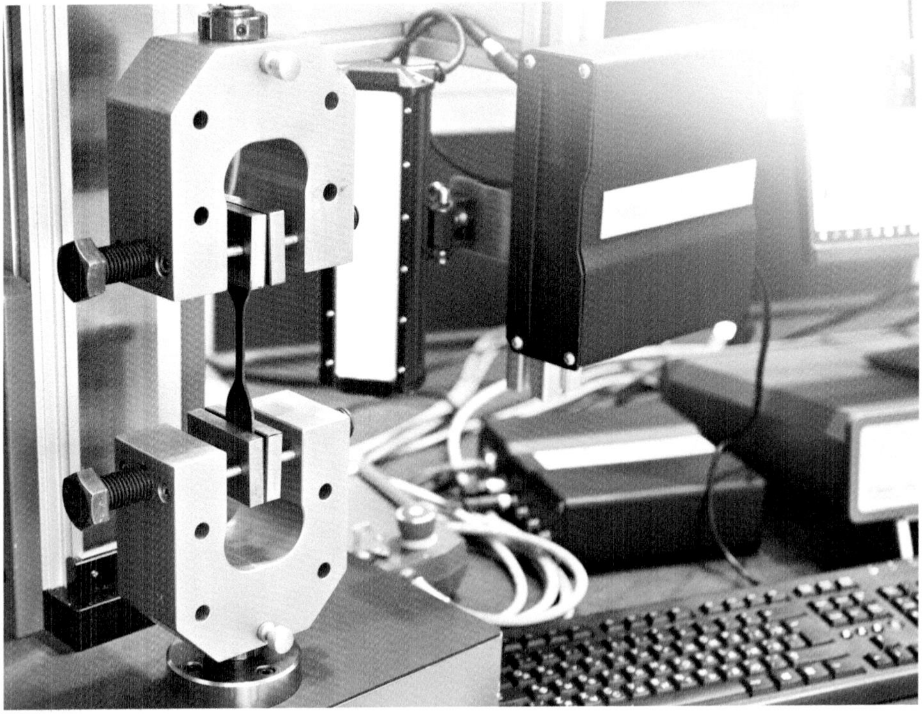

(a)

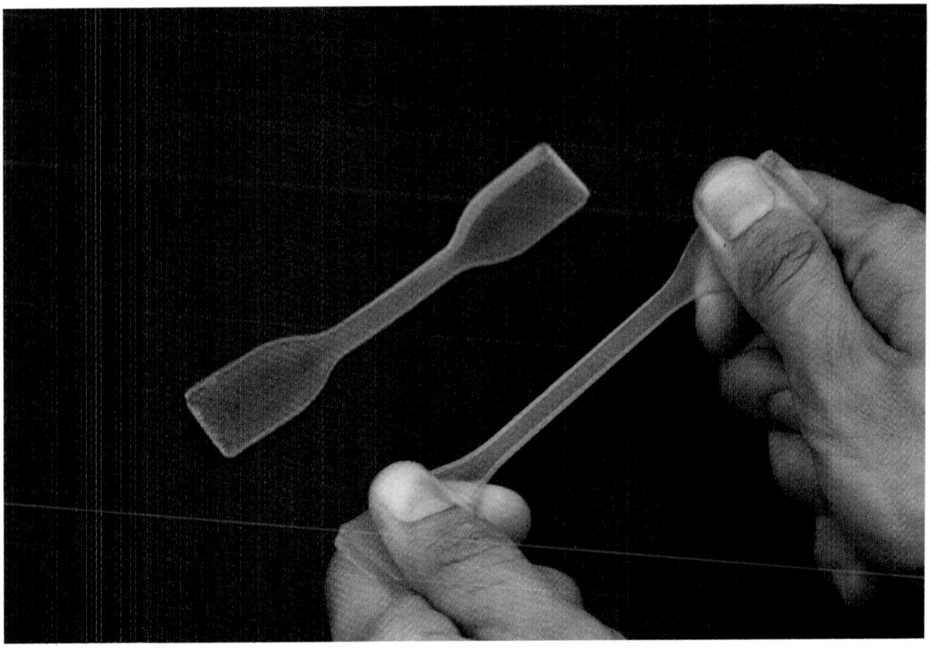

(b)

Figure 5.6 Photos showing examples of (a) tensile sample in a universal test device and (b) tensile test samples.

Figure 5.7 Photo showing a three-point flexural test with a universal test device.

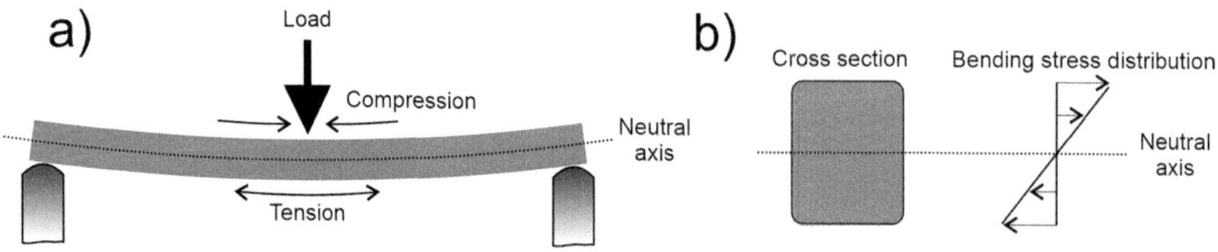

Figure 5.8 Illustration of (a) a rectangular cuboid sample in a three-point bend test configuration, showing compressive strain on the side of the load and tensile strain on the opposite side, and (b) the bending stress distribution of the beam in (a).

axis (Figure 5.8). The upper half of the sample (side being pushed down) experiences compression, while the lower half (side on supports) experiences tension. For truly homogenous materials with equal compressive and tensile moduli, the properties obtained from a flexural test will match those from a compression or tensile test. Most material samples contain defects and imperfections, which cause differences between tensile and compressive response. This means properties often differ between uniaxial and flexural tests.

5.15 Considerations when characterising sports materials

5.15.1 Strain measurement

It can be beneficial to measure the strain of a sample directly. There are two main strain measurement approaches, (i) contact, e.g., clip-on extensometer or strain gauge, and (ii) optical, e.g., laser or camera, including digital image correlation [138]. Contact strain measurement techniques are accurate, but they are typically discrete, meaning the measurement/s are limited to the location of the attached sensors or device. Contact with the sample (i.e. gluing a gauge to its surface) can also influence its behaviour. In contrast, optical strain measurement techniques, particularly digital image correlation, offer "full-field" measurement. The lack of contact means these techniques do not influence sample behaviour. For example, optical full-field strain measurement techniques are useful for measuring Poisson's ratio (Equation 5.6).

5.15.2 Stress softening and the Mullins effect

Some materials, particularly rubbers, will have a unique stress–strain response the first few times they are loaded to a given strain. After about five loading cycles the stress–strain curve will "settle" and be consistent for subsequent loadings. This is the process of stress softening, or the Mullins effect [139]. It must be considered when characterising rubber-like materials, such as those used in tennis balls. One option is to cyclically load the sample about five times, or until stress–strain curves become consistent, and then use results for subsequent cycles.

5.15.3 Loading rate

The behaviour of some materials will depend on the rate at which they are loaded, or the strain rate. Given the dynamic nature of sport, it is important to consider the strain rate when characterising sports materials. Think of those used in sports balls, protective equipment, or running shoes, for example. Impact tests, and dynamic mechanical test devices, can be used to produce the high strain rates typical of sporting scenarios [140, 141].

5.16 Mechanics of materials

5.16.1 Flexural rigidity

The flexural stiffness of an item of sports equipment, such as a tennis racket, ski, baseball bat, or pole vault pole, is dependent on both its construction material and shape. To make calculations easier and more efficient, such items are often assumed to act as uniform cross-section beams [28]. The flexural rigidity or bending stiffness (EI) of such a beam is the product of the elastic modulus of the material and the second moment of area (I) of its cross-section (i.e., shape). The second moment of area for a solid rectangular cross-section about the neutral axis is:

$$I_r = \frac{bd^3}{12} \qquad \text{Equation 5.12}$$

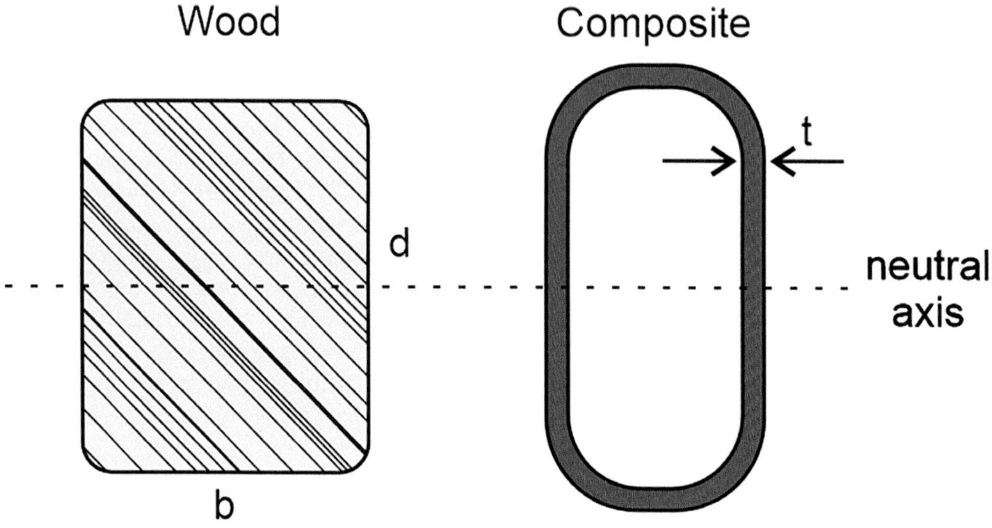

Figure 5.9 Illustration showing basic approximations of typical cross-section shapes for traditional wooden and modern composite tennis racket frames, showing the width (b), depth (d), and wall thickness (t). These are just indicative, and the specific shape of the cross-section often varies around the racket frame.

where b is width and d is depth (Figure 5.9). From Equation 5.12, the most efficient way (i.e., least material) to increase the second moment of area, and hence flexural rigidity, of a rectangular cross-section beam is to increase its depth (i.e., thickness in the direction of bending) rather than its width.

The second moment of area for a hollow rectangular cross-section about the neutral axis is:

$$I_{r_h} = \frac{\left[bd^3 - (b-2t)(d-2t)^3\right]}{12} \qquad \text{Equation 5.13}$$

where t is wall thickness (Figure 5.9). Essentially, for a hollow rectangle, the second moment of area of the hollow inner section is subtracted from the solid outer one.

Using a hollow cross-section is an effective way of reducing the mass of a beam-like structure without substantially reducing its flexural stiffness. This is because material located further from the neutral axis has a larger effect on the second moment of area. Striking implements, such as tennis rackets, were traditionally wooden with solid cross-sections, and the transition to composite materials allowed them to be made hollow (e.g., Figure 5.9) [28, 30]. Racket cross-sections became deeper (in the direction perpendicular to the string-bed) to maximise out-of-plane bending stiffness. The ability to make hollow cross-sections with composites contributed to modern rackets being lighter and stiffer than their wooden predecessors.

The second moment of area of a solid circular cross-section is defined as:

$$I_c = \frac{\pi}{4} r^4 \qquad \text{Equation 5.14}$$

where r is radius. The second moment of area of a hollow circular cross-section is defined as:

$$I_{c_h} = \frac{\pi}{4}\left(r_o^4 - r_i^4\right) \qquad \text{Equation 5.15}$$

where the subscripts o and i correspond to the outer and inner radii, respectively. There are various resources covering the second moment of area of other shapes (e.g., [87]).

Consider a pole vault pole with a hollow circular cross-section. The pole's second moment of area can be tailored to the desired flexural rigidity by adjusting the inner and outer radius along its length. Similar concepts can be applied to bicycle frames, which can be considered as a collection of hollow tubes, specifically designed to give desirable stiffness (see Figure 5.10). Developments in materials and manufacturing techniques, particularly composites engineering, have allowed sports engineers to vary the cross-sectional shape throughout bike frames to give desirable riding characteristics for different disciplines.

5.16.2 Bending stress in beams

Bending stress is:

$$\sigma_b = \frac{My}{I} \qquad \text{Equation 5.16}$$

where M is moment (see Equation 3.4), and y is the distance from the neutral axis. Equation 5.16 tells us that the bending stress in the cross-section of a beam subject to bending increases with the distance from the neutral axis (Figure 5.8). The bending stress is zero at the neutral axis and highest on the upper and lower surface. One side of the beam will be in tension and the other compression.

Bending stress increases with the distance from the neutral axis because the strain also increases (see Equation 5.3). The scenario becomes more complex if the beam is heterogeneous (e.g., sandwich construction), e.g., a surfboard with a lightweight foam core and stiff composite skin. The stiff skin on the outside of the board will stretch or compress (strain) more under bending than the foam core because it is further from the neutral axis. This means the stiff skin will have a larger bending stress. These principles apply to many beam-like sporting goods, such as rackets, bats, boards, and skis and pole vault poles. The value of the bending stress determines whether such structures will perform adequately or if they will fail. Failure will occur if the bending stress exceeds the failure criteria of the material, as already described. The specific bending stress distribution will depend on the construction materials and whether they are homogenous or heterogeneous, e.g., sandwich construction or fibre-matrix composites (see ref. [142]).

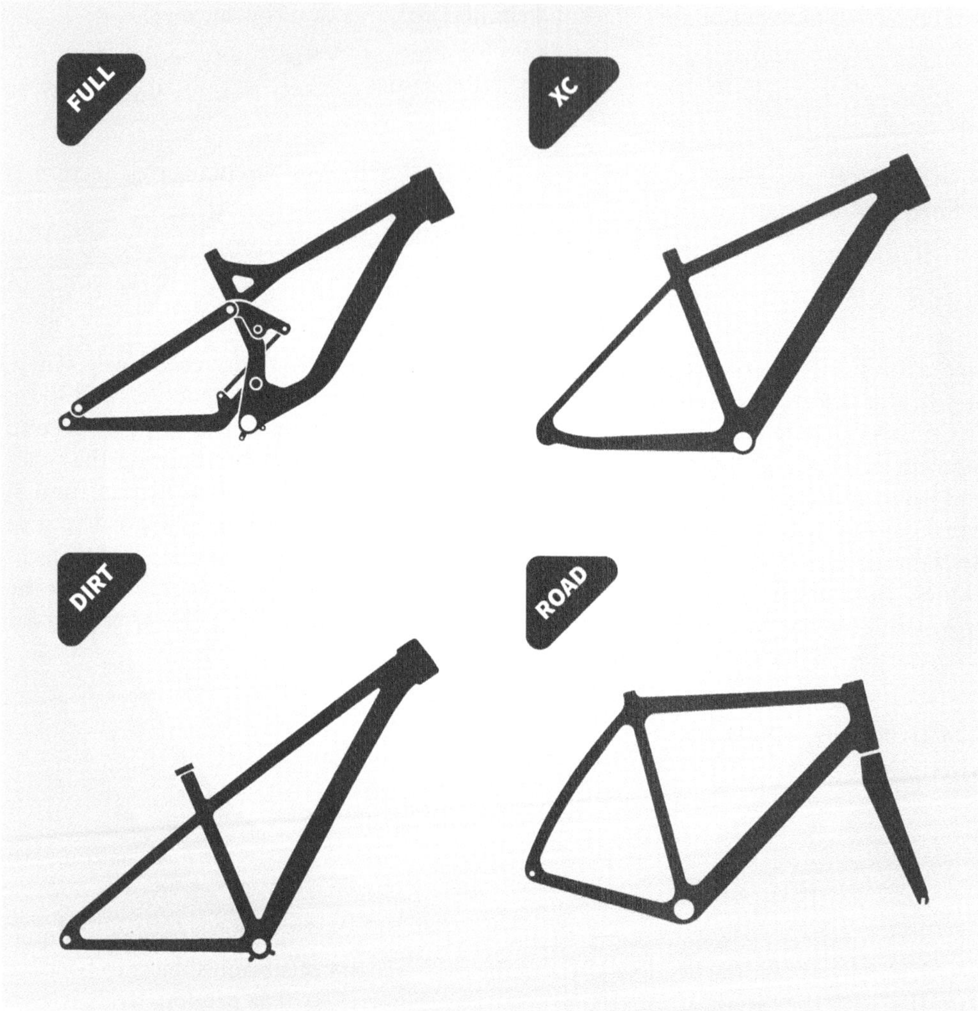

Figure 5.10 Illustration showing typical shapes of bike frames for different purposes, including full suspension/downhill, cross country/hard tail, dirt jumping, and road riding. Differences in the shape and size of the tubes, both within and between bikes, can clearly be seen. These differences influence the stiffness of the frame.

5.16.3 Buckling

A tall, narrow column will buckle under a sufficiently high compressive axial load. Instability from buckling causes lateral deflection (Figure 5.11a). Euler buckling theory defines the critical load that will cause an ideal column to buckle as:

$$F_c = \frac{\pi^2 EI}{(kL)^2} \qquad \text{Equation 5.17}$$

where k is a factor related to the end conditions and L is the column's length. Ideal columns have a uniform cross-section and are homogenous and isotropic. Such columns will buckle if the critical load is exceeded.

How easily a column buckles (i.e., F_c) depends on the end conditions. For example, the factor k equals 1 when both ends of the column are pinned and 0.5 when they are fixed. Effective length (l_e) is the product of the factor k and column length (i.e., kL), which allows simplification of Equation 5.17 to:

$$F_c = \frac{\pi^2 EI}{l_e^2} \quad \text{Equation 5.18}$$

The theory shows that longer columns with lower flexural rigidity (EI) buckle more easily. Such columns have a low slenderness ratio (λ), or slenderness, which is:

$$\lambda = \frac{l_e}{k} \quad \text{Equation 5.19}$$

where k is the radius of gyration:

$$k = \sqrt{\left(\frac{I}{A}\right)} \quad \text{Equation 5.20}$$

and where A is cross-sectional area. Columns with small cross-sectional dimensions have a low radius of gyration and hence can have a high slenderness ratio (Equation 5.19), meaning they buckle easily (see Figure 5.11b).

By comparing the critical buckling stress to the yield stress of the material, we can determine if a column will fail under compression (i.e., crushing) or buckling. First, Equation 5.20 can be rearranged to make second moment of area (I) the subject:

$$I = Ak^2 \quad \text{Equation 5.21}$$

and then substituted into Equation 5.18:

$$F_c = \frac{\pi^2 EAk^2}{l_e^2} = \frac{\pi^2 EA}{\left(\frac{l_e}{k}\right)^2} \quad \text{Equation 5.22}$$

and further simplified using Equation 5.19 to:

$$F_c = \frac{\pi^2 EA}{\lambda^2} \quad \text{Equation 5.23}$$

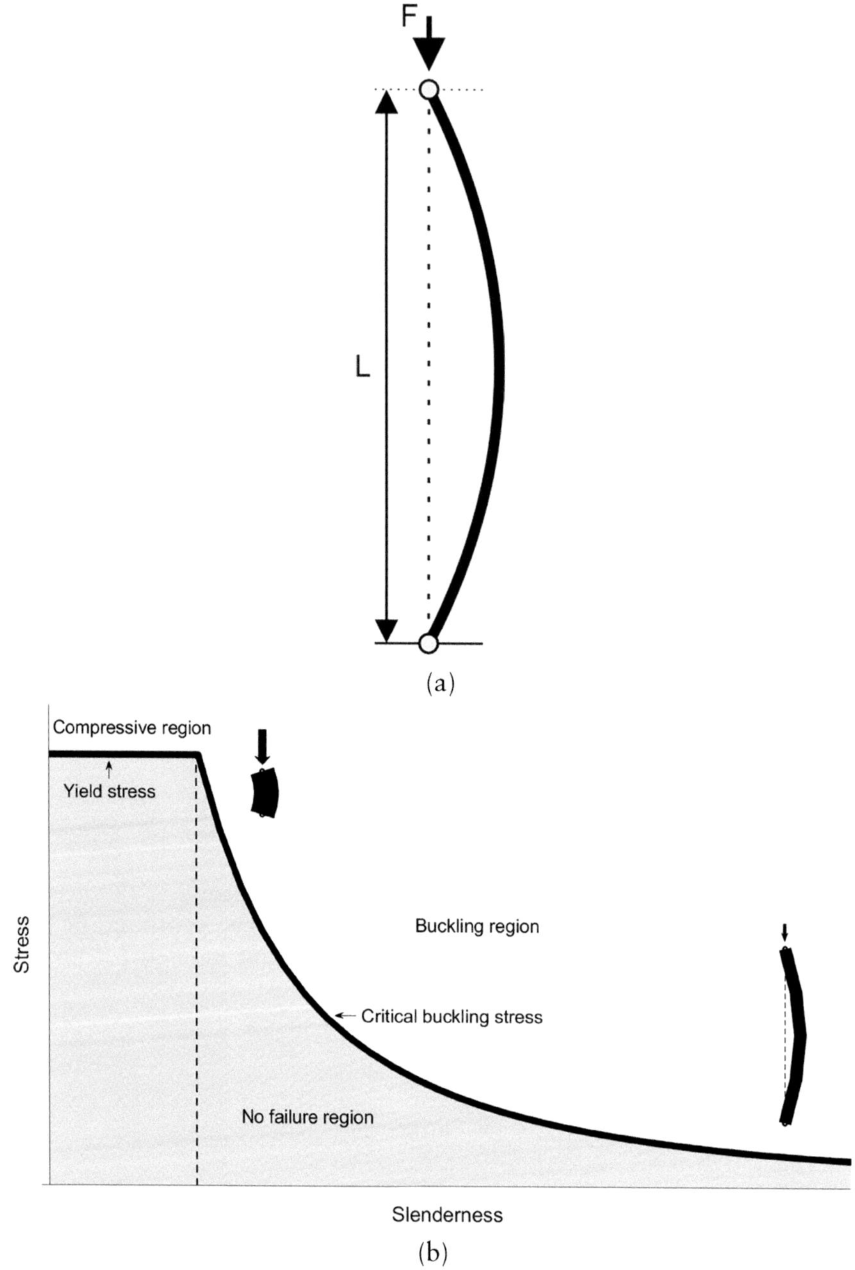

Figure 5.11 (a) Illustration of a column that is pinned at both ends buckling under an axial load (F), and (b) a plot to illustrating the theoretical relationship between the stress and slenderness of a column, according to Euler buckling theory. The "no failure" grey region below the thick line corresponds to scenarios where the applied stress is below the failure threshold for a column. The region above the thick line corresponds to scenarios where the applied stress is above the failure threshold for a column, meaning it will fail via compression or buckling. The curved region of the line shows that the critical buckling stress increases as a column becomes less slender, i.e., it becomes harder to buckle (compare images of stocky beam on left and slender beam on right). If the slenderness of the column is sufficiently low for the theoretical critical buckling stress to become larger than the yield stress of the material, it will fail under compression (i.e., crushing) rather than buckling.

As stress is the ratio of force to cross-sectional area (see Equation 5.1), the critical buckling stress is:

$$\sigma_c = \frac{F_c}{A} = \frac{\pi^2 E}{\lambda^2} \qquad \text{Equation 5.24}$$

If the critical buckling stress is lower than the yield stress, the column will fail via buckling (see Figure 5.11b).

The Euler buckling theory has limitations in practical use, including not accounting for imperfections in actual columns. The concepts can, however, still be applied to various sporting goods, such as sports balls, where the walls buckle under compressive impacts [143], and the wheels and frames of bikes. For example, large-diameter hollow tubes are used in bike frames. This is partly because their large diameter gives them a high second moment of area, resulting in a low slenderness ratio and good buckling resistance, while the hollow construction keeps weight low. The theory can also be applied to metamaterials structures, which can be designed to buckle under specific conditions to give the desired properties, e.g., auxetic. Metamaterials are covered in Chapter 6.

Given the complexity of sporting goods, in terms of materials and geometry, finite element analysis is often used for structural analysis and design (e.g., Figure 5.12). A detailed description of the technique is outside the scope of this textbook. There are many books, learning resources, and research articles on finite element modelling (e.g., [10, 144–149]). A good place to start is the website of the International Association for the Engineering Modelling, Analysis and Simulation Community.[2]

Figure 5.12 Image showing an example of finite element analysis of a bike frame, showing Von Mises stress.

5.16.4 *Fatigue and failure*

Fatigue is a common failure mechanism for sports equipment and components, such as bike frames and components, skis and boards, and buckles. Fatigue is the, typically gradual, process by which a material or component degrades and weakens with repeated loading, often leading to failure. Repeated loading below the failure criteria for one loading cycle can cause micro damage and micro cracks, which gradually grow and lead to failure. Factors such as material and manufacturing imperfections, sites of damage (e.g., rock impact on a bike part), stress concentrations (often from poor design), and corrosion can all accelerate the effects of fatigue.

As with the single load-cycle failure, the fatigue behaviour of a material or component is typically determined empirically. That is by an experimental process of trial and error. For example, consider a bike handlebar in a three-point bend test configuration. The first step would be to load the handlebar to the point of failure, to determine its failure load for one loading. Once this failure load is known, the number of loading cycles required to cause failure of identical samples at a lower load can be determined. For example, a handlebar could be cyclically loaded to 90% of the one cycle failure load until failure, to determine the number of cycles to cause failure at 90% of its one cycle failure load. The process could then be repeated at 80%, 70%, 60%, etc. of its one cycle failure load, allowing generation of a plot of failure load vs. the number of cycles. When testing material samples, such a plot of failure stress vs. the number of cycles is typically referred to as an S–N plot (e.g., Figure 5.13). This type of fatigue analysis is important for materials and sports equipment that will be repeatedly loaded, such as mountain bikes, skis, and ice hockey sticks.

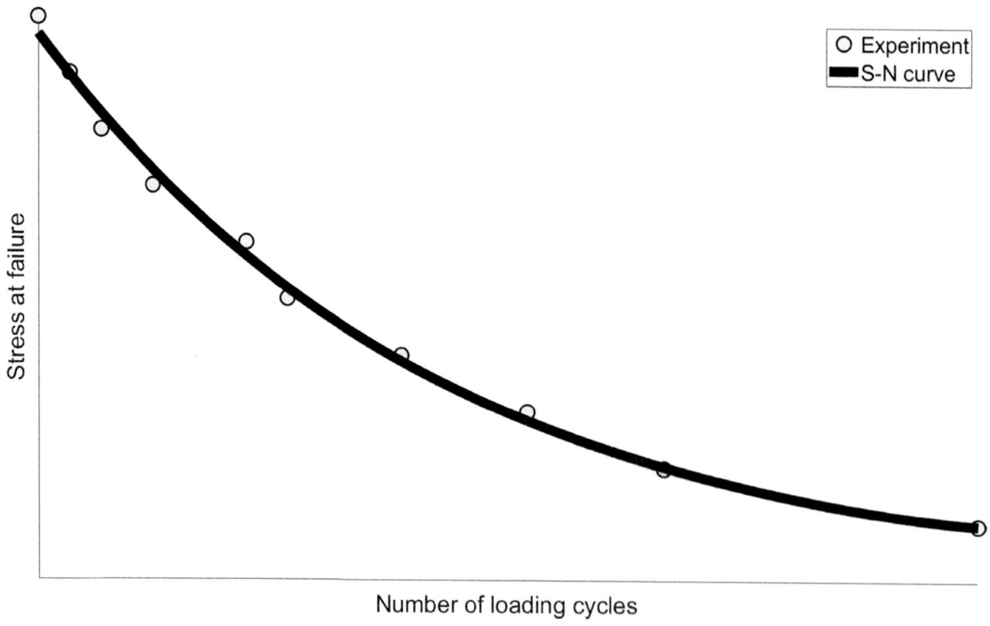

Figure 5.13 Illustration of an S–N curve, illustrating the relationship between the number of loading cycles and the stress at failure. The experimental data and curve are just indicative to illustrate the concept. The shape of an actual S–N curve will depend on the specific material.

Materials fail by growth of internal cracks, rather than direct separation of chemical bonds. This means we often need to use combined stresses, rather than stress in one direction, to predict failure risk. These internal cracks tend to open when shear stresses are high and are often more resistant to volumetric stress, so failure criteria often use functions of shear strain energy. These failure criteria include Von Mises, which is often used for ductile materials, and Tresca, which is more conservative and often used for brittle materials (see [87, 150]). In general, highly brittle materials fail in tension, as cracks open, so these combined stresses are not needed.

5.17 Chapter summary

This chapter has provided a broad overview of sporting materials. It covered the fundamentals of materials and how they relate to sporting goods. This basic understanding should be sufficient to see how materials relate to the complementary topics covered in this introductory textbook on sports engineering. There is, however, still more to be learnt about materials beyond the basics covered here. Students are encouraged to continue reading and learning about materials. The next chapter will go into more detail about how materials are applied to sporting goods.

5.18 Suggested further reading

The following suggestions for further reading go deeper into the topics introduced in this chapter:

- *Engineers' Data Book*, 4th edition [87]
- *Mechanics of Materials*, 10th edition [150]
- *Materials in Sports Equipment*, 1st and 2nd editions [133, 134]
- *Materials Selection in Mechanical Design*, 4th edition [130]

Notes

1 Co-authored by Dr Olly Duncan of Manchester Metropolitan University.
2 https://www.nafems.org/

6 Materials in sporting goods[1]

Building on the previous chapter, this one focuses on the use of materials in sporting goods. It highlights the different types of materials used in sporting goods. These materials can be categorised in various ways, and the six broad categories shown in Figure 6.1 are used here. There is some overlap between these categories, which span both "soft" (e.g., apparel) and "hard" (e.g., equipment) sporting goods. The chapter concludes by highlighting the new and emerging material concepts being applied and tested in sporting arenas, including those that are non-Newtonian or have negative or otherwise counterintuitive properties. A recent development that is covered is the emerging field of metamaterials, which have potential to improve sports equipment and enhance safety [22, 23].

6.1 Natural materials

Sporting goods were traditionally made from natural materials, particularly wood. This subsection focuses on wood. Wood was often used by early sports engineers because it was readily available and easy to craft. There are two broad categories of wood, softwood and hardwood, with the latter typically being denser, stronger, and more durable. While wood is still used for some sporting goods, in many cases it has been surpassed by manmade alternatives. For example, wood is now rarely used in rackets, bikes, paddles, oars, and pole vault poles, but it is still used in skateboard decks, hurling sticks (hurley) and table tennis, cricket, and baseball bats (e.g., Figure 6.2). Wood is also used in sports floors, e.g., for basketball.

There are various reasons why the use of wood has either persisted or been surpassed, which are often related to the specifics of the sport. For some sports, such as cricket, it is about preserving traditions. While wood has some advantages, it has characteristics that can cause design challenges, as typical of many natural materials [151].

Some advantages are:

- Readily available
- Several varieties with different properties
- Can be crafted into different shapes
- Can be layered as a plywood with fibres in different directions to tune properties
- Buoyant
- Tends to resist failure by preventing cracks from propagating (see ref. [151]).

DOI: 10.4324/9781003411086-6

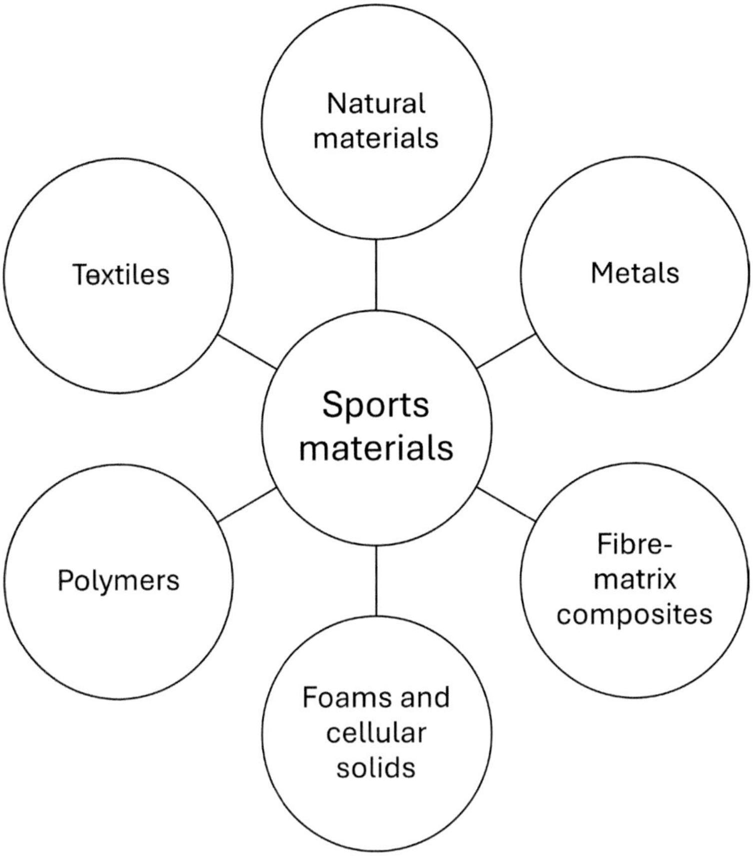

Figure 6.1 Illustration showing the broad categories of sports materials used in this textbook.

Some characteristics that can cause design challenges, and that apply to other natural materials, are:

- Inherently variable and inconsistent, both within and between samples
- Heterogenous and anisotropic
- Sensitive to moisture
- Can fail unpredictably

For many sporting goods brands, the mechanical and manufacturing constraints of wood motivated them to seek alternative materials, often metals, polymers, and composites. Governing body rules restricting the use of other materials have often been the reason why certain sporting goods are still made from wood, as is the case with cricket and baseball bats. With concerns over climate change driving a move

(a)

(b)

Figure 6.2 Photos showing examples of wooden sporting goods, (a) skateboard deck, (b) hurley, (c) cricket bat, (d) baseball bat, and (e) bamboo bike frame (note bamboo is a grass).

(c)

(d)

Figure 6.2 (Continued)

(e)

Figure 6.2 (Continued)

towards more sustainable materials, those that are natural, including wood, may see a resurgence in sporting goods. For example, bamboo has been used for bike frame tubes (e.g., Figure 6.2).

6.2 Metals

The low cost and desirable properties of metals (see Figure 5.3 and Table 5.1) have allowed them to find various applications in sporting goods. Examples include ice skate blades, golf clubs, gym weights, climbing and mountaineering equipment, helmet face shields, the edges of skis and snowboards, runners for luges, skeletons or bobsleighs, skateboard trucks, and bike frames and components (e.g., Figure 6.3). Metals are particularly suitable for bike frames and components like seat posts, stems, and handlebars. They can be formed into lightweight hollow tubes with a high flexural rigidity (*EI*), and bonded at the ends, often via welding, to form a bike frame (see Chapter 7). The commonly used metal alloys, steel, and aluminium, and to some extent titanium, differ sufficiently to allow sports engineers to develop bikes with different properties and characteristics, to suit individual riding styles and preferences. Interestingly, the specific stiffness (see subsection 5.6) is similar for all three of these alloys (~25 $Mm^2/$

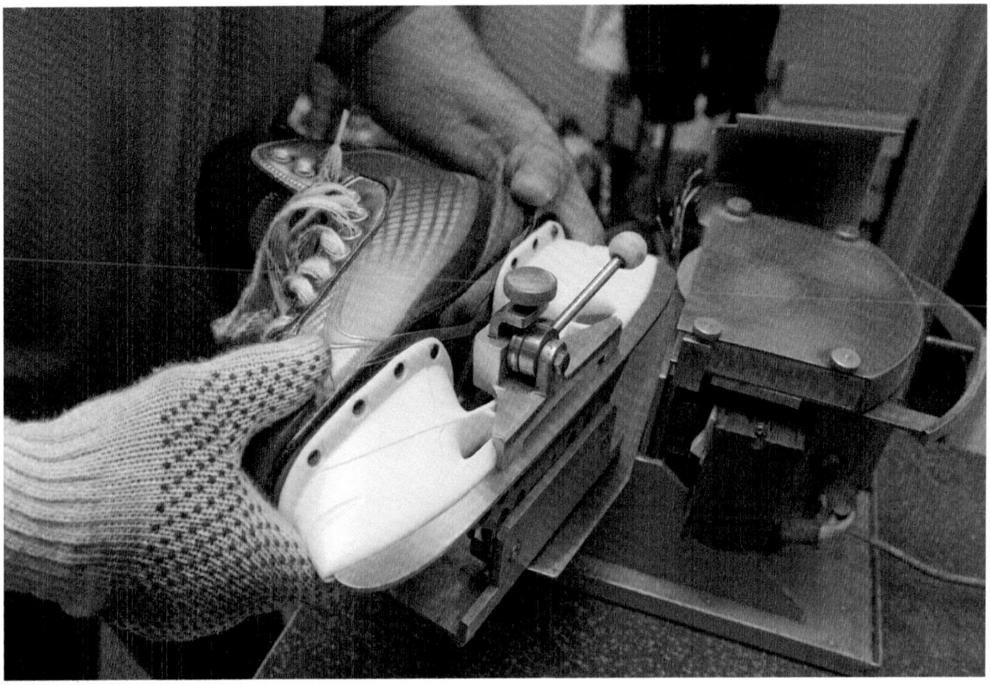

(a)

(b)

Figure 6.3 Photos showing examples of metal in sporting goods, (a) ice skate blades, (b) ski edges, (c) mountaineering equipment (ice axes, karabiners, crampons), (d) dumbbells, and (e) bike frame and components.

(c)

(d)

Figure 6.3 (Continued)

(e)

Figure 6.3 (Continued)

s^2, Table 5.1). Via changes in the alloy, cross-section, and length of the tubes, various bikes of different size and riding characteristics can be produced efficiently (see Figure 5.10).

For many sports, metal has, or is, being replaced by fibre-matrix composites. This is mainly due to the superior properties of composites (see Figure 5.3 and Table 5.1, specific modulus and specific strength), and the extra design freedom they offer. For example, while traditional metal bike frames are triangulated for strength, the adoption of fibre-matrix composites has permitted the use of non-traditional and more organic forms in cycling. There can also be marketing benefits of using materials that may be perceived as being "advanced" or "exotic", such as titanium and composite materials. Environmental concerns, associated with end-of-life recycling of composites, and their extra cost, combined with individual user preferences, mean metals will likely continue to have a place in sporting goods.

6.3 Fibre-matrix composites

Fibre-matrix composites, or composites, are widely used in sporting goods. They combine fibres with high tensile stiffness and strength with a polymer matrix. Fibres commonly used in sporting goods are E-glass, carbon fibre, and aramid. The diameter of these fibres is typically 8–12 μm [133]. A key advantage of such composite materials is the design freedom they offer to sports engineers. For example, their properties can be altered by varying the material, volume fraction (i.e., percentage), orientation, and placement of the fibres. The ratio of fibres to matrix is typically 0.5~0.75 (i.e., 50–75% fibres), depending on the application.

Unidirectional fibre-matrix composites have all the fibres aligned in one direction. For these composites, the "rule of mixtures" can be used to determine key properties based on those of the constituent materials and their volume fraction. The density of the composite (ρ_c) is:

$$\rho_c = V_f \rho_f + V_m \rho_m \qquad \text{Equation 6.1}$$

where the subscripts *c*, *f*, and *m* correspond to the composite, fibre, and matrix, respectively. Similarly, the composite's Poisson's ratio (υ_c) and Young's modulus (E_c) parallel to the fibres are:

$$\upsilon_c = V_f \upsilon_f + V_m \upsilon_m \qquad \text{Equation 6.2}$$

and

$$\mathrm{E}_c = V_f E_f + V_m E_m \qquad \text{Equation 6.3}$$

Imagine a unidirectional fibre-matrix composite consisting of an equal volume of fibres and matrix. In this example, the density, as well as the Poisson's ratio and Young's modulus (when loaded parallel to the fibres), are simply the mean of the corresponding values for the fibre and matrix (Table 6.1). If the fibre volume fraction was increased from 0.5 to 0.75, then the properties of the composite become closer to those of the fibre (denser and stiffer in this example) (Table 6.1).

Unidirectional fibre-matrix composites are anisotropic, so their mechanical response depends on the loading direction. The Young's modulus normal to the fibres in unidirectional composites is much lower than the modulus parallel to the fibres. As such, bidirectional fibre-matrix composites are often used to achieve more uniform biaxial properties. The calculations to obtain the biaxial properties of such composites are more complex than those described above, and are covered elsewhere, e.g., [134, 148]. The mechanical properties of fibre-matrix composites and structures can also be obtained using computer-aided engineering software, e.g., finite element analysis software.

The various applications of composites include boats, pole vault poles, running blades, rackets, skis and snowboards, and high-end bikes (see Figure 6.4). Various constituent

Table 6.1 Example properties (parallel to fibres) for hypothetical fibre-matrix materials consisting of different ratios of fibres and matrix. These properties are just indicative and are intended solely as an example of rule of mixture calculations.

Material	*Density (kg/m³)*	*Poisson's ratio*	*Young's modulus (GPa)*
Matrix	1.2	0.30	3
Fibre	2.0	0.20	250
Composite – 50% fibres	1.6	0.25	127
Composite – 75% fibres	1.8	0.23	188

(a)

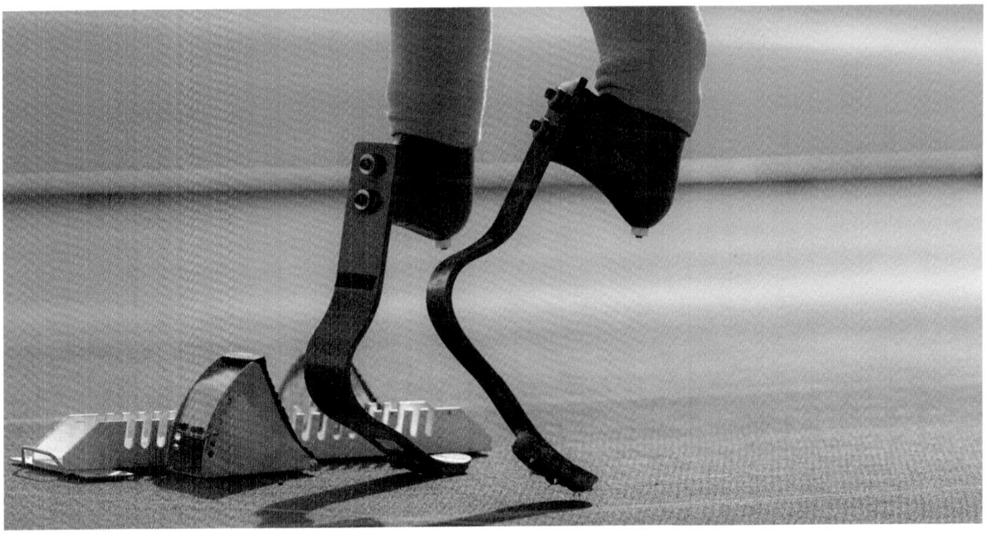

(b)

Figure 6.4 Photos showing examples of composites in sporting goods, (a) tennis racket, (b) running blades, and (c) bike.

(c)

Figure 6.4 (Continued)

materials can be combined and positioned in different locations and orientations within such products. Composites also allow for sports products where the shape, and wall thickness, vary throughout. This ability to locally adjust the material, fibre-matrix orientation, thickness, and shape gives engineers the freedom to create a diverse range of sporting goods with characteristics tailored to different applications and end users, e.g., a tennis racket or bike frame where the material properties and cross-sectional shape vary throughout the product to give the required stiffness. Composite bike frames and components can be made in streamlined shapes to reduce aerodynamic drag (see Figure 6.4c).

Despite the advantages of composites, as with other manmade materials, there are environmental concerns associated with their use in sporting goods [15]. It is challenging, and energy intensive, to separate the constituent materials at the end of the product's life for recycling. Manufacturing processes for composite sports products also tend to be labour-intensive, as it is difficult, and hence costly, to automate production. Manufacturing of composites is covered in Chapter 7.

6.4 Foams and cellular solids

There are various types of cellular solids, including foams, honeycombs, and lattices. Polymeric foams are widely used in products like fitness and crash mats [152], surfboards, running shoes, and protective equipment, including in body padding and helmets (Figure 6.5). The properties of foams are well suited to such applications. They are lightweight and have distinct compression characteristics (Figure 6.6). There are three main types of foam, open-cell, closed-cell, and expanded. As air can freely pass through open-cell foams, they tend to be lighter and softer than their closed-cell counterparts. Expanded foam is formed by initially expanding pellets, with steam or hot air, followed by further expanding and bonding via steaming and cooling in a mould. For example, helmet liners are often moulded in expanded foam (Figure 6.5d, [22]). The properties of a foam can be tuned by changing the material and the relative density (Figure 6.6).

The compressive stress–strain curve of a foam has three main parts (Figure 6.6). The first is a relatively stiff linear region up to 5~10% compression. The second is a low-stiffness plateau region, where the cell walls buckle. The third is densification, where the cell walls "self-contact", the constituent material compresses, and stiffness increases. If densification is reached during an impact, this is typically referred to as "bottoming-out". The area under the curve is the energy absorbed per unit volume, i.e., strain energy from compressing the foam.

An effective protective device keeps stress (i.e., force) below a harmful level (i.e., injury threshold). It typically does this by absorbing impact energy while keeping force low, i.e., without densifying and bottoming out, so the force stays below the injury threshold. The plateau region allows energy to be absorbed with minimal change in stress. Foam that is too stiff gives a high impact force because this plateau region is never reached. Foam

(a)

Figure 6.5 Photos showing examples of foam in sporting goods, (a) fitness mat and roller, (b) surfboard core, (c) running shoe midsole, and (d) cross-section of ice hockey helmets.

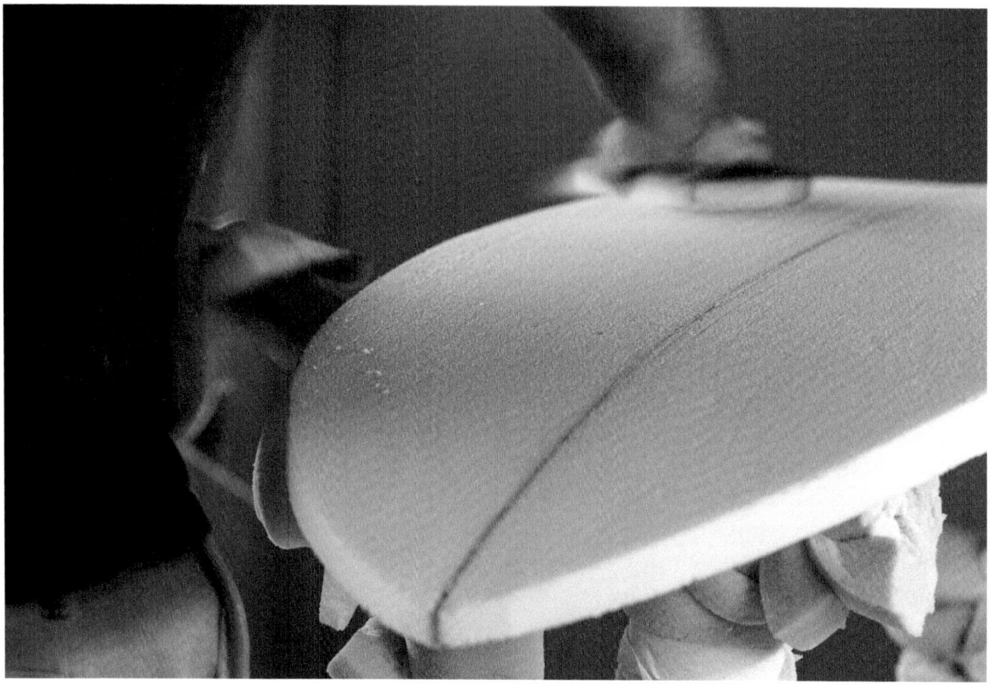

(b)

(c)

Figure 6.5 (Continued)

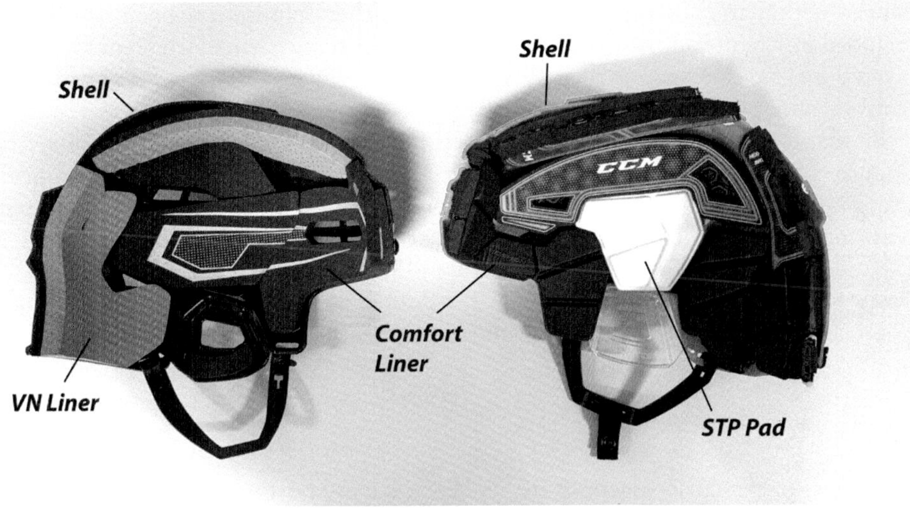

(d)

Figure 6.5 (Continued)

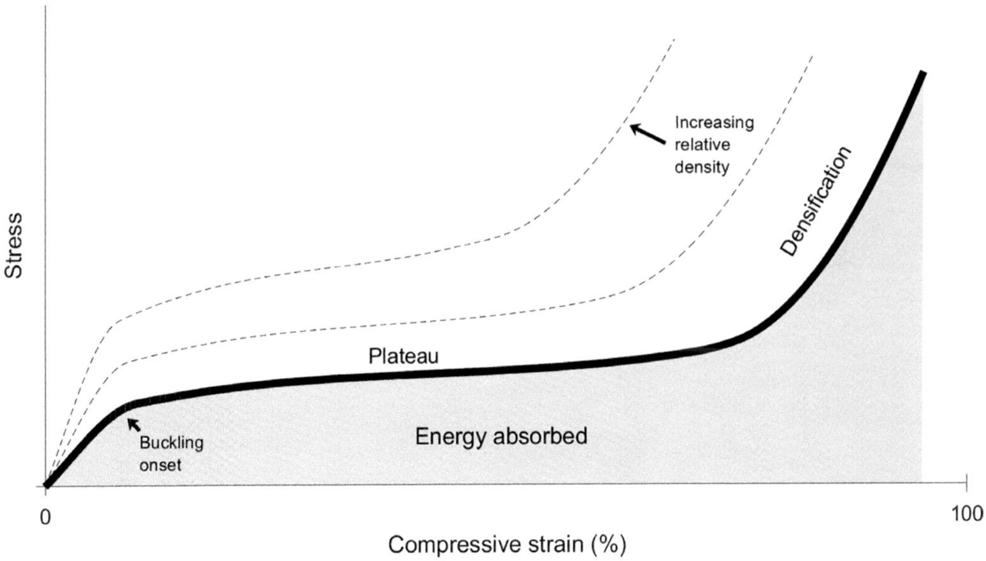

Figure 6.6 Illustrative plot showing a typical compressive stress–strain curve for foam. Annotations show the onset of buckling at the end of the initial high stiffness region, the plateau region, and the densification region. The area under the stress–strain curve represents the strain energy density. The effect of increasing the relative density of the foam is also shown.

that is too soft will reach densification under impact, which also causes high force. The properties of the foam and the size (particularly thickness) of the device can be tuned to the intended impact scenario to ensure compression does not surpass the plateau region and cause densification. This often means designing the device to perform well in tests defined within any corresponding standards. Further details on the role of standards are included in Chapter 8. Foams with different properties, particularly density, can also be layered in a sandwich construction, to expand the range of impacts over which the protective device performs well.

Similar principles apply to the closed-cell foams used in running shoe midsoles (Figure 6.5b) (see refs. [42, 153, 154]). In this use case, lightweight and resilient (high energy return) foam can help the wearer run more efficiently. Shoe brands often work with materials suppliers to develop and market specific foams for their products. The expanded foam cores of surfboards (Figure 6.5b) are encased within a fibre-matrix composite skin, typically fibreglass, to give the desired stiffness and strength (see Chapter 5). For readers seeking more detail, cellular solids are well documented in the literature (e.g., [10, 155]).

6.5 Polymers

Polymers, particularly plastics, are widely used in sporting goods [156]. The various examples include sports balls, racket strings, protective equipment shells, skateboard wheels, ski boots and bindings, bike components, and casings for wearable devices (see Figure 6.7). A noteworthy example is the introduction of polyurethane skateboard wheels in the 1970s, which improved the riding experience and helped popularise the sport. There are various types of polymers, and different properties can be obtained by adjusting the blend (mixture of at least two polymers) and production process. As well as

(a)

Figure 6.7 Photos showing examples of polymers in sporting goods, (a) ski boots, (b) snowboard bindings, (c) skateboard wheels, (d) bike components, and (e) casing of wearable device.

(b)

(c)

Figure 6.7 (Continued)

(d)

(e)

Figure 6.7 (Continued)

having desirable mechanical properties (Figure 5.3 and Table 5.1), polymers are generally low-cost, readily available in different colours, easy to mould into different shapes and sizes, and non-corrosive. This is why polymers are widely used in sporting goods.

Sports engineers interested in using polymers must appreciate the difference between thermoplastic (or thermo-softening) and thermosetting (Figure 6.8) [156]. Thermoplastic polymers soften when heated and can be remoulded and recycled. Thermosetting polymers do not soften when heated. They are durable with good temperature and chemical resistance but are hard to recycle.

There are increasing concerns about the environmental impact of polymers, particularly around their end-of-life disposal in landfill [15]. This is a particular issue for those that are bonded or moulded to other materials, making them hard to separate and recycle. As such, sports engineers are investigating alternatives to polymers. They are also exploring ways to reduce the number of polymer types within a given product to help facilitate recycling. Nonetheless, polymers remain efficient to produce, durable (allowing long product lifespans), and can often be recycled if isolated from other materials.

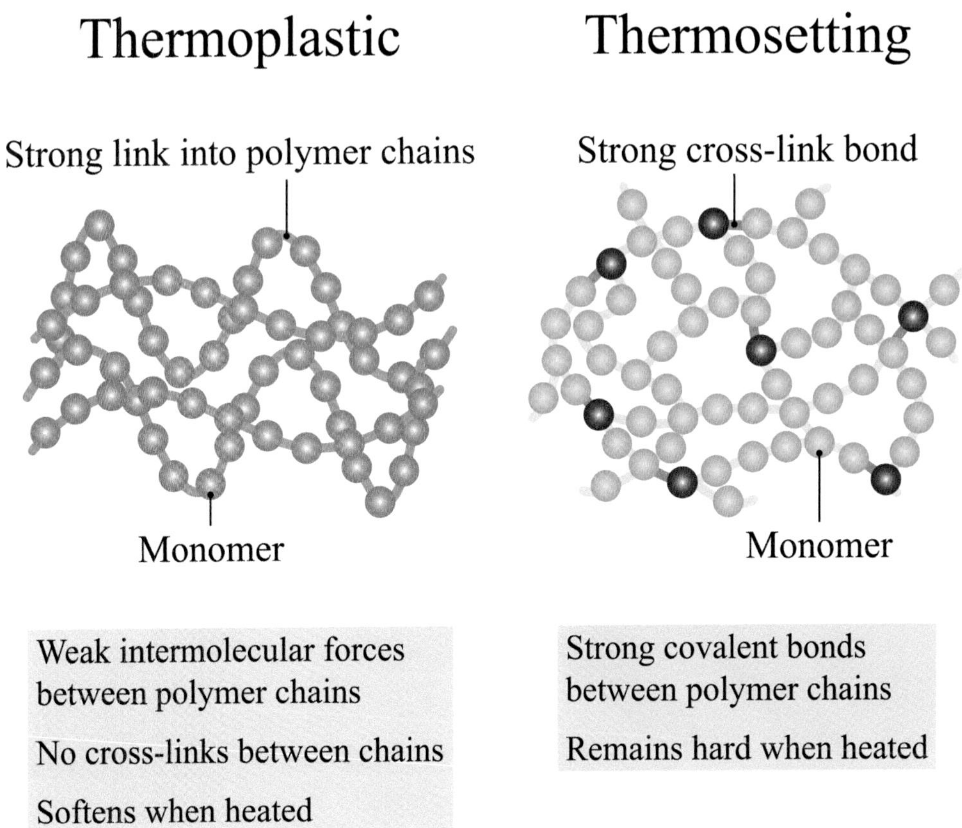

Figure 6.8 Illustration showing the difference between thermoplastic and thermosetting. Thermoplastics can be remelted, while thermosetting plastics remain in a solid state.

Removal of polymers from sports products must be done carefully, while considering lifecycle analysis, to limit the risk of replacement with a less sustainable material.

The properties of polymers, particularly thermoplastics, can be sensitive to changes in temperature. For example, some polymers are stiff at room temperature, while others are flexible and rubber-like. The latter are a sub-set of polymers called elastomers. They are highly elastic, meaning they can experience high strains, and still return to their original size, i.e., without plastic deformation. Elastomers are useful for sports products like balls and soft components like handlebar grips and snowboard binding straps. How a polymer behaves at different temperatures is related to its glass-transition temperature. Below this temperature a polymer will be harder and more brittle (i.e., glass-like). As the temperature increases past the transition, the polymer will become softer and more rubber-like. It is important for sports engineers to understand how a polymer will behave across different temperature ranges, particularly for winter sports equipment.

6.5.1 Textiles

Textiles are widely used in sporting goods, and typically those categorised as "soft goods". The various sporting applications of textiles include clothing, footwear, bags, ropes, tents, nets, wingsuits and parachutes, and sails (see Figure 6.9). Textile structures are made from textile fibres. These fibres can be natural (from animals or plants) or man-made (synthetic), each with their own advantages. The properties of a textile structure depend on how it was constructed and the base textile fibre properties.

(a)

Figure 6.9 Photos showing examples of textiles in sporting goods, (a) running gear, (b) outdoor gear, (c) a tent, (d) wingsuits, (e) snow sports gear, (f) climbing gear, and (g) a volleyball net.

(b)

(c)

Figure 6.9 (Continued)

(d)

(e)

Figure 6.9 (Continued)

(f)

(g)

Figure 6.9 (Continued)

There are three main types of textile structures, woven, knitted, and non-woven. The first two types are created by interlacing or interlooping yarns via weaving or knitting. Yarns are continuous strands of fibres or filaments that are twisted together. Non-woven textiles are produced by bonding textile fibres directly, typically via a mechanical, chemical, or thermal process. Different textile fibres, including natural and synthetic, can be blended in a textile structure, to tailor their properties and characteristics for specific sporting applications. Electronics can also be integrated into textile structures (at the yarn, fabric, or finishing stages) to create "smart textiles" (E-textiles). These smart textiles include those that can collect data, such as temperature or heartrate, from the wearer and provide feedback via a device such as a smartphone.

As with other materials used in sporting goods, there are environmental concerns around textile use in mass-produced products (e.g., clothing), and the associated "fast fashion" throwaway culture [15]. This is particularly true when multiple fibre compositions, accessories, and electronics are combined, as this complicates recycling processes. As such, sports engineers are taking steps to minimise the environmental impact of soft goods. These steps include exploring the use of natural alternatives to synthetic fibres, reducing the variety of fibre materials within a given product, increasing manufacturing efficiency (e.g., using offcuts), developing recycling processes and practices (e.g., customers returning goods at the end of their life for recycling), and pursuing circular design to support a circular economy.

6.6 New and emerging material concepts in sporting goods

This subsection concludes the chapter by highlighting new and emerging material concepts being applied and tested in sporting arenas. Such materials typically have unusual or counterintuitive properties. The topics covered are non-Newtonian materials and metamaterials.

6.6.1 *Non-Newtonian materials*

Viscosity, as introduced in Chapter 4, is a material's resistance to flow (see Figure 6.10a). Fluids with low viscosity, such as water, flow more easily than those with high viscosity, such as honey. Viscosity tends to be influenced by temperature. For example, honey flows more easily when warm than when cold, due to a decrease in its viscosity.

Newtonian materials have an almost linear relationship between shear stress and shear strain rate (see Figure 6.10b). This means that the viscosity of Newtonian materials barely changes with shear strain rate (see Figure 6.10b). Non-Newtonian materials have a non-linear relationship between shear stress and shear strain rate (see Figure 6.10b). This means that their viscosity changes with shear strain rate (see Figure 6.10b). These materials are sometimes referred to as strain rate sensitive. Two common types of non-Newtonian materials with potential applications in sporting goods are shear thinning and shear thickening materials. The viscosity of shear thinning materials decreases with shear strain rate (see Figure 6.10 [tomato ketchup is a well-known example, which is why a lot falls out of the bottle when you hit it]). The viscosity of shear thickening materials increases with shear strain rate (see Figure 6.10b). Shear thickening materials are sometimes called dilatant.

Shear thickening materials are generally more applicable to sporting protective devices, as they become harder to shear with increasing shear strain rate. They have found application in personal protective equipment, such as footwear and gloves, and knee, elbow, and back protectors. This is because, in simple terms, shear thickening materials

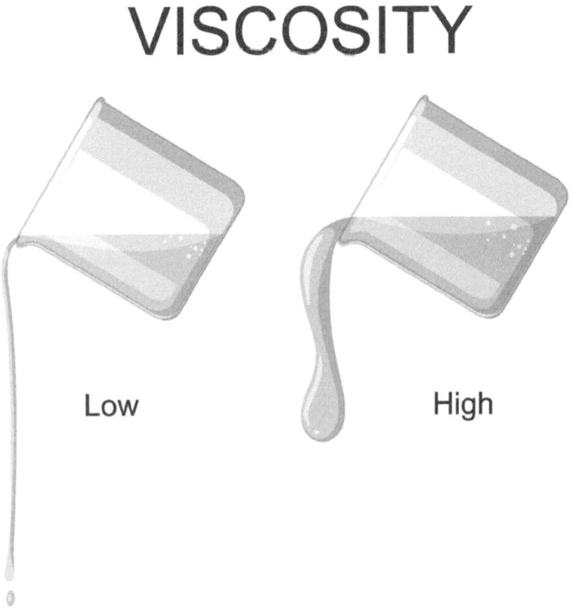

(a)

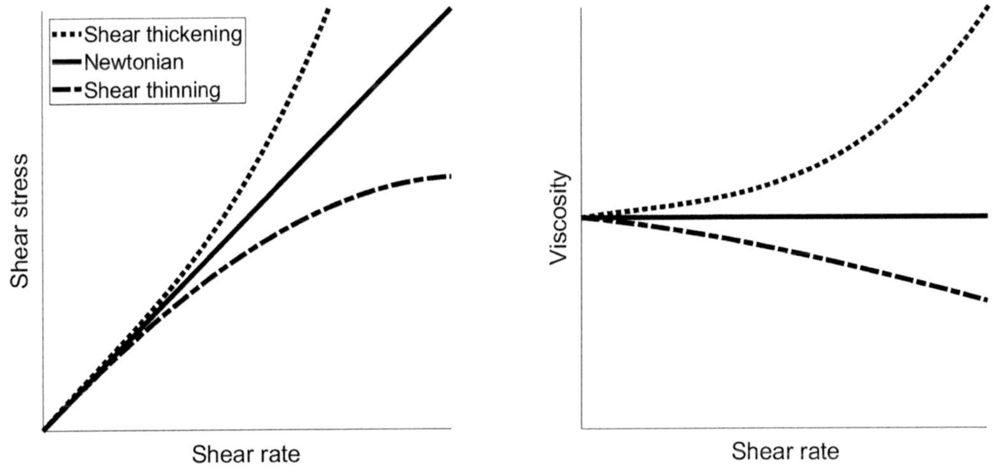

(b)

Figure 6.10 (a) Illustration showing liquids with low and high viscosity, respectively, and (b) graphs illustrating the relationship between (left) shear stress and shear rate and (right) viscosity and shear rate for shear thickening, Newtonian, and shear thinning materials.

can be soft and flexible under normal use but hard and stiff under impact to enhance protection (see Chapter 2 and Chapter 8). Shear thickening materials can also offer enhanced vibration damping. A well-known example of a shear thickening material is the toy Silly Putty™, which can be slowly deformed by hand but bounces when dropped onto a solid surface.

The two main types of shear thickening materials are fluids (suspensions) and gels. Shear thickening fluids tend to be less stable, and hence harder to work with, than shear thickening gels. They also need to be enclosed (e.g., within a plastic pouch) for use in products. As such, shear thickening gels tend to be more suitable for application in sporting goods. The crosslinked polymer chains within shear thickening gels make them more viscous than their fluid counterparts, and hence more stable and easier to work with. These materials are often made by combining boron and silicone, with the specific ratio influencing properties such as viscosity.

Shear thickening gels can be combined with other materials, such as textiles and foams, to produce samples with shear rate-dependent properties [22, 157, 158]. For academic research, this is often via a secondary process, such as coating the material with the shear thickening fluid, absorbing (e.g., soaking) it into the material, or sandwiching it between layers of the materials (e.g., between sheets of foam). It is also possible to 3D print shear thickening materials, and then enclose them in an elastic material, such as silicone, so they can recover their shape after being deformed [159]. Processes more suited for mass production have been developed for commercialising sporting goods with shear thickening materials (see Figure 6.11) [24], e.g., injection moulding or making foam with shear thickening material.

As well as footwear and body armour, shear thickening materials are being used in sports helmets, with the intention of reducing concussion risk from impacts that cause the head to rotate (e.g., [160]). A common limitation of shear thickening materials is their temperature-sensitive properties, which is a key consideration for winter sports. More can be done to further our understanding of shear thickening materials and to tune their properties specifically for application in sporting goods. They also have potential to enhance metamaterial design and application in sporting goods.

6.6.2 *Mechanical metamaterials*

This section summarises metamaterials and their current and potential application in sporting goods. Metamaterials are engineered structures with unique and unusual properties that cannot usually be obtained from the materials they are made from. These properties are often counterintuitive, e.g., moulding a flexible polymer in a specific geometric pattern that will cause it to fatten laterally when stretched longitudinally. As already noted, such a metamaterial would be auxetic, as it would have a negative Poisson's ratio (Equation 5.6). Mechanical metamaterials have found more applications in sporting goods than the other types, so they will be the focus of this section.

Mechanical metamaterials can be made in various ways from conventional (bulk) materials. They are typically made by cutting, moulding, or additively manufacturing a material to create a structure with a repeating pattern. The units of repeated geometry that form the structure are often referred to as unit cells. The specific design of the unit cells and how they are patterned determine the properties of the metamaterial.

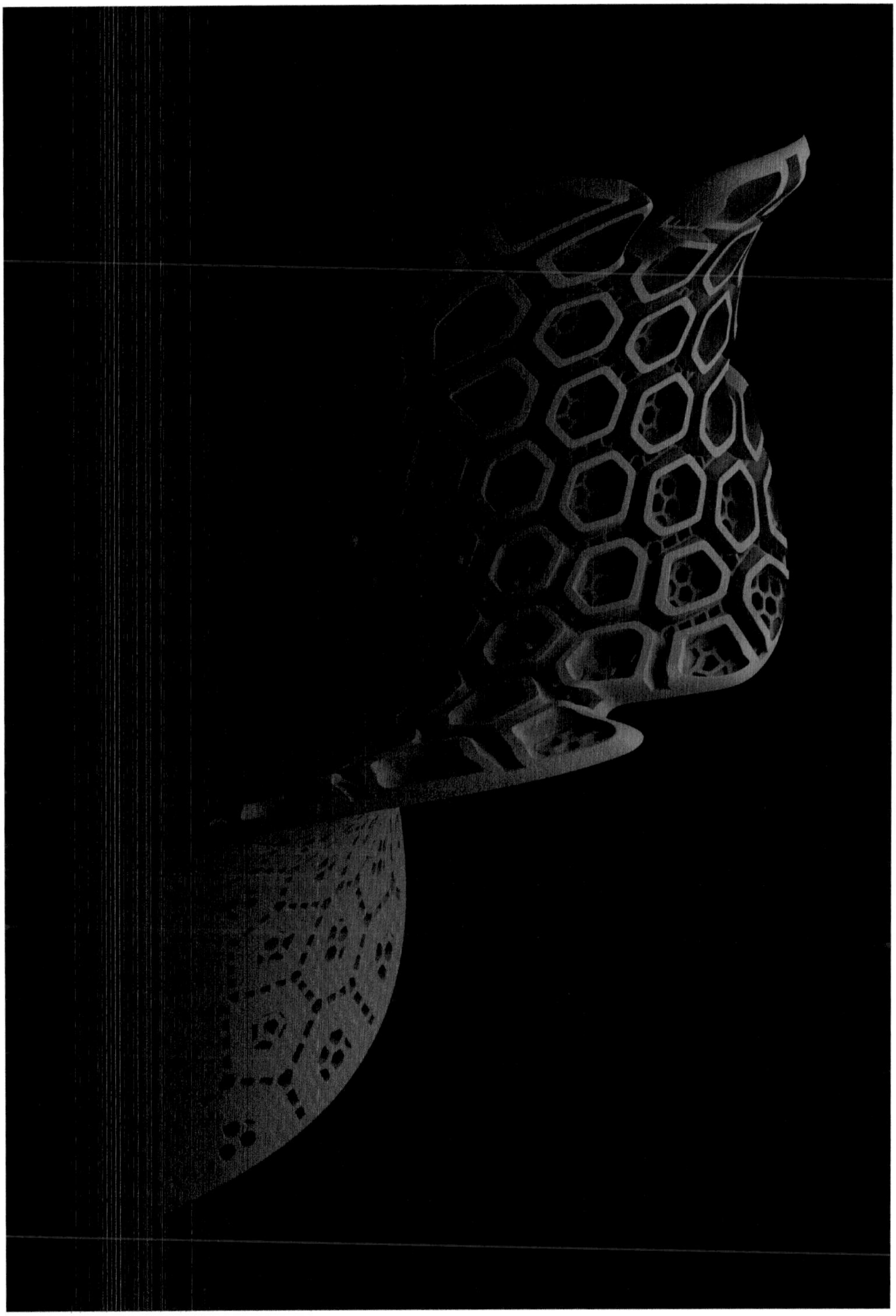

Figure 6.11 Photo showing an example of body protection containing shear thickening material. The flexibility of the product under low strain rates is demonstrated via twisting. Image courtesy of Rheon Labs Ltd.

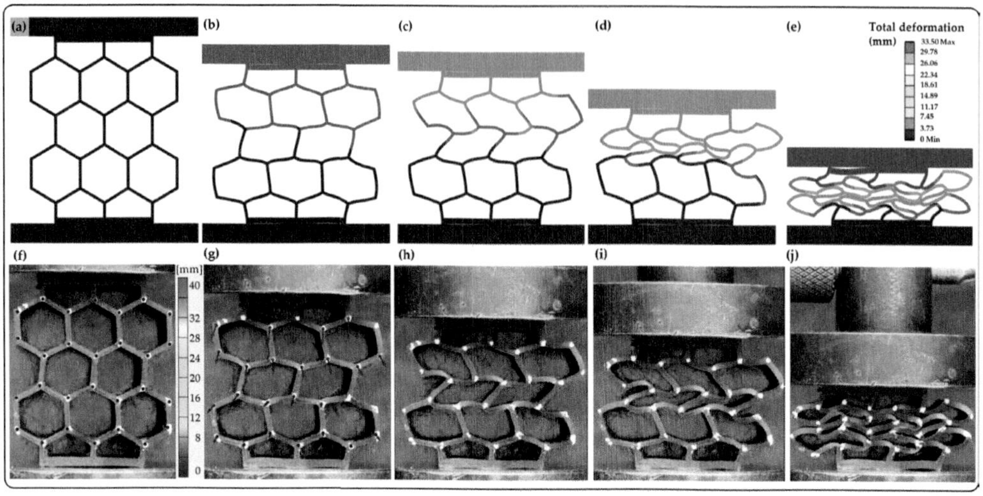

Figure 6.12 Image sequence showing an example of a metamaterial style honeycomb being tested under compression, (top) virtually in finite element modelling software and (bottom) physically as a 3D printed sample. Image from ref. [161]. Licensed under CC BY 4.0.

Two-dimensional extruded structures (honeycombs) are simpler to make than three-dimensional structures (lattices), particularly via cutting or moulding.

The emergence of additive manufacturing has helped progress work in mechanical metamaterials, and the two topics complement each other. For example, additive manufacturing has helped researchers to produce and explore lattices, including those made from multiple materials to enhance their properties. This can include a difference in material stiffness between the ribs and junctions of unit cells. Finite element modelling is also a useful design tool for virtually testing and refining mechanical metamaterials (e.g., Figure 6.12 [161]). The following three types of mechanical metamaterials will be covered in this sub-section:

- Auxetic
- Negative stiffness
- Adaptive

6.6.3 *Auxetic*

Characteristics of auxetic materials that could bring benefits to sporting goods include enhanced vibration damping, high indentation resistance (hardness) and synclastic (domed) curvature (Figure 6.13) [22, 23]. Such curvature has been utilised in products like body protection and helmet liners, with the intention of improving fit. The lateral expansion of auxetic materials under tensile load could also enhance the comfort of wearable items, e.g., by allowing the straps of sports bras and rucksacks to expand under load and distribute pressure more evenly to the wearer.

The force required to push a hard shape into an elastic half space is:

$$F = \mathrm{D}\left(\frac{E}{1-v^2}\right) \qquad \text{Equation 6.4}$$

Figure 6.13 Photos showcasing how (a) conventional and (b) auxetic samples curve under an out-of-plane moment. The auxetic sample has the established re-entrant, bowtie-shaped, structure. Image from ref. [23]. Licensed under CC BY 4.0.

where the constant D relates to the size, shape, and depth of the indentation [162]. As Poisson's ratio tends towards the lower limit of –1 for isotropic materials, the denominator in Equation 6.4 tends towards zero and hence infinite indentation resistance is theoretically possible. In other words, no matter the size or shape of the indenter, an idealised auxetic material (with $v = -1$) will deform as though it is being compressed by a flat plate. This means auxetic body protection could be soft and flexible under normal use, while also resisting concentrated impact loads (e.g., falls/collisions with rocks and tree roots during mountain biking). This is somewhat like the application of shear thickening materials in protective devices, and attempts have been made to combine the two concepts [157].

Fibre-matrix composites can also be designed to be auxetic via specific fibre alignment [22, 23, 163]. This means that such composites can be made with the same constituent materials and tools as their conventional counterparts, as it is just the fibre alignment that needs adjusting. This low barrier to entry has helped auxetic composites find application in tennis and padel rackets, and they will likely soon be used in other sporting goods. Auxetic composites can be more resistant to back face damage under impact. As such, they have potential to enhance the durability and lifespan of products like body armour, bikes, snowboards, and skis.

6.6.4 Negative stiffness

Negative stiffness can be achieved by including "snap-through" elements in a metamaterial [22] (Figure 6.14). One example of such an element is a beam in a buckled configuration (Figure 6.14). Negative stiffness can be achieved by applying a lateral load against the direction that the beam is curving during buckling (left region of plot in Figure 6.14). If this lateral load is large enough, the beam will enter an unstable region (central region of plot in Figure 6.14) before snapping through to stabilise in its reflected buckling orientation (right region of plot in Figure 6.14). The unstable snap-through region has negative stiffness, because the force decreases as the displacement increases (i.e., while it snaps through).

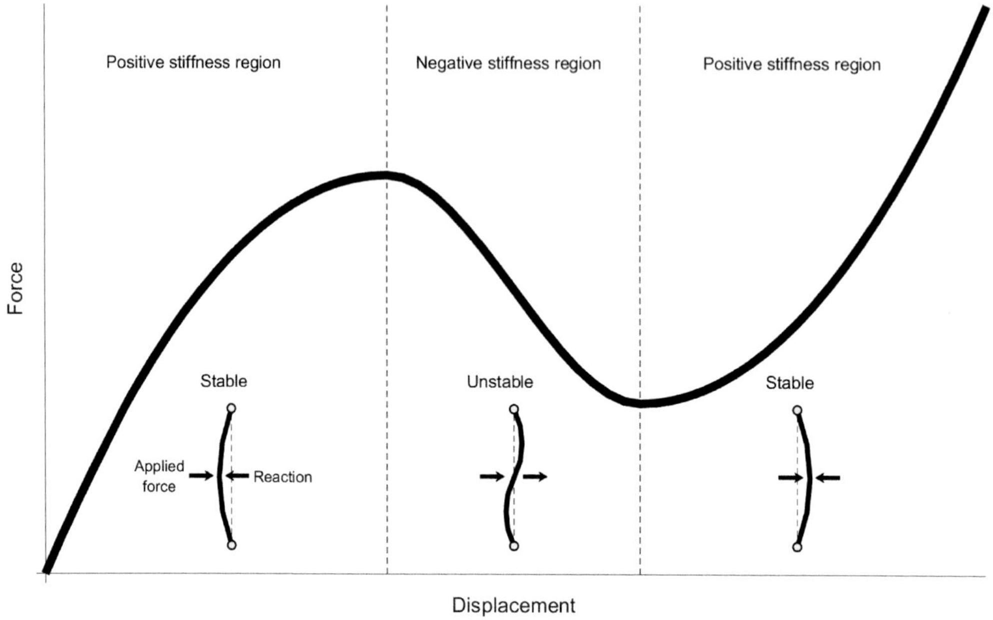

Figure 6.14 Illustration of a force vs. displacement plot for a negative stiffness element, in the form a buckled beam loaded laterally at its centre. Inserts show the beam in its initial stable configuration (buckled to left), its intermediate unstable configuration as it "snaps through", and its final stable configuration (buckled to right).

The shape of the force–displacement relationship depends on the properties of the beam. For example, a slenderer beam (longer and narrower) will snap through more easily and will have lower positive and lower negative stiffness. Negative stiffness behaviour could enhance the vibration and impact performance of sporting goods. For negative stiffness metamaterials to gain widespread uptake in sporting goods, they must be easily mass-producible. Reference [164] presents a relatively simple negative stiffness metamaterial design with potential for mass production.

6.6.5 Adaptive metamaterials

Adaptive metamaterials are an emerging class of material that could bring benefits to sporting goods. As the properties of mechanical metamaterials depend on their structure, changing the shape of this structure will change their properties. There are various ways in which the shape of a metamaterial structure could be changed. One way is to actively force a change in the shape of the structure. This could, for example, be achieved using embedded electromagnets, by changing the tension of embedded cables/fibres, or by changing the air pressure within cavities. Another way is to change the loading response of the material the metamaterial is made from. This could, for example, be achieved using materials with strain rate sensitive properties (see prior subsection [6.6.1] on non-Newtonian materials). Another option could be using conductive polymers that change their stiffness under an electric charge [165]. These concepts will now be explained further via the example of a bi-beam (two laterally connected beams of different stiffness) [166].

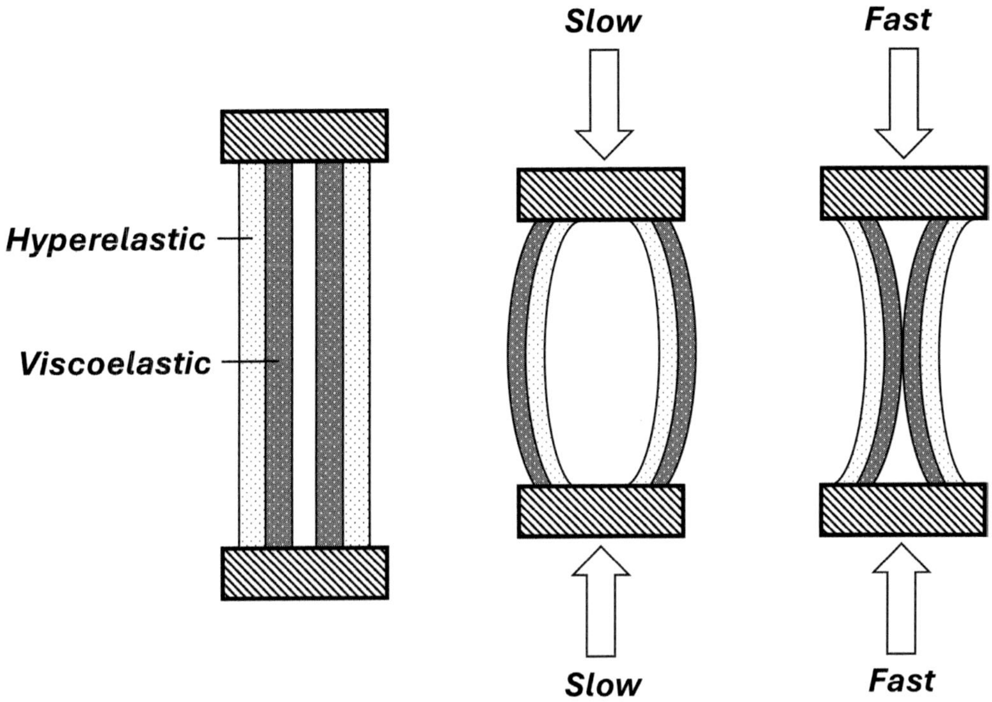

(a)

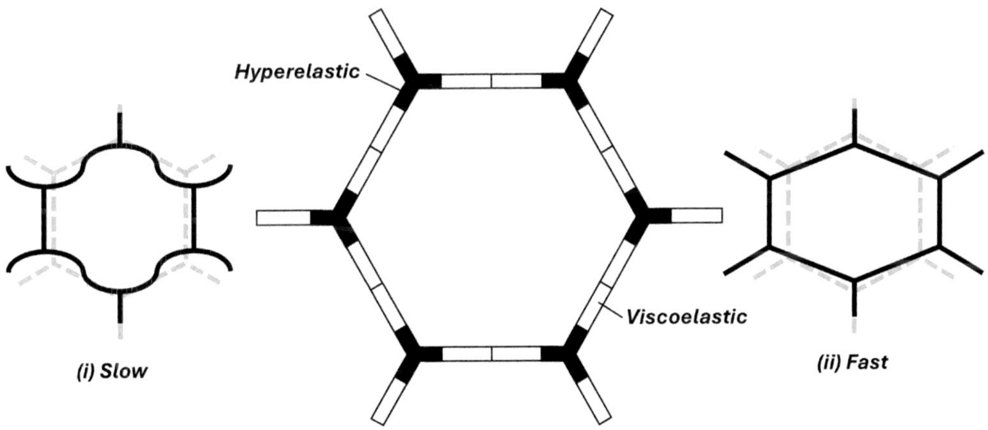

(b)

Figure 6.15 Diagrams showing examples of mechanical metamaterials designed to have strain rate-dependent properties, (a) a single bi-beam design and buckling direction for different axial compressive strain rates, and (b) multi-material hexagonal unit cell causing a switch in the dominant deformation mode at different strain rates. Based on figure 7 of ref. [22].

As already described in Chapter 5, a slender beam will buckle when compressed axially (see Figure 5.11). The direction of lateral deflection will be towards the beam's stiffer side. For a supposedly uniform beam, imperfections in materials and manufacturing can influence the buckling direction. For a bi-beam, the lateral deflection will be towards the stiffer of the two laterally connected beams. So, by changing which of these paired beams is stiffer, the direction of lateral deflection can also be changed. For example, if each beam was made of a piezoelectric material that stiffens when charged, the buckling direction could be changed by charging one of the beams. Another option is to have a bi-beam that changes its buckling direction with the compression rate. This could be achieved by using a strain rate-sensitive material for one of the paired beams that is less stiff than its partner when loaded slowly and stiffer when loaded quickly. Another option could be to pair strain softening (shear thinning) and strain stiffening (shear thickening) beams.

By placing such bi-beams in parallel to form unit cells, the concept of an adaptive metamaterial could be realised (see Figure 6.15a, [22]) [166], e.g., a metamaterial that switches between conventional and auxetic behaviour, or vice versa, as compression rate increases. Similar concepts can also be applied to different unit cell design shapes and structures (see Figure 6.15b, [22]) to create various adaptive metamaterials, e.g., by changing the relative stiffness between ribs and junctions. The unique and unusual properties of such adaptive metamaterials could then be controlled for different sporting applications and scenarios. This could include apparel, braces (e.g., wrists, knees, neck) and other wearable items that adapt to the fit and needs of the wearer, or bikes, boards, and skis that adapt to the terrain and riding scenario. While adaptive metamaterials clearly have potential for enhancing sporting goods, they are an emerging topic that requires further research before application in consumer products can be realised. The various challenges to their application range from taking them beyond initial concepts to better understanding their design and properties to the cost and environmental considerations of mass production.

6.7 Chapter summary

This chapter has provided a broad overview of the application of materials in sporting goods. It finished by summarising some new and emerging concepts of sports materials. Readers should now understand how materials are used in sporting goods. This understanding should be sufficient to see how materials relate to the complementary topics covered in this introductory textbook on sports engineering. There is, however, still more to be learnt about the application of materials in sporting goods beyond the basics covered here. Students are encouraged to continue reading and learning about materials and how they can, and will, be applied in the sporting goods of today and tomorrow.

6.8 Suggested further reading

The following suggestions for further reading go deeper into the topics introduced in this chapter:

- "Contemporary advances in polymer applications for sporting goods: fundamentals, properties, and applications" [156]
- *Materials and Technology for Sportswear and Performance Apparel* [167]

- "Mechanical metamaterials for sports helmets: structural mechanics, design optimisation, and performance" [22]
- "Review of auxetic materials for sports applications: expanding options in comfort and protection" [23]
- *Routledge Handbook of Sports Technology and Engineering*, 1st and 2nd editions [10, 86]

Note

1 Co-authored by Dr Olly Duncan of Manchester Metropolitan University.

7 Manufacturing sporting goods

Sporting goods manufacturing is a multi-billion-dollar industry. It ranges from daily leisurewear (i.e., "athleisure") and apparel for mass participation, to highly specialised and custom-created bespoke equipment for high-performance sport. Sporting goods manufacturing is therefore a broad landscape, with diverse processes and technologies employed through various price points. The industry is large and fast-paced, driven by ever-changing user requirements, technological innovation, regulatory updates, trends in fashion and marketing, and global sporting events.

The diverse needs of users present unique challenges within sporting goods manufacture. This means manufacturing choices must balance for various factors, including equipment function, the scale of a given market, the product's performance requirements, and cost. Accordingly, material selection plays a key role in manufacturing processes' selection. A comprehensive understanding of material selection and processing is a crucial element of equipment design and manufacturing. Almost all manufacturing processes can be found within the sporting goods industry, but it is dominated by large-volume production of products with relatively short lifecycles [168]. This includes "fast fashion" items, which often end their life in landfills [15]. Thankfully, sustainability considerations are becoming increasingly important to consumers. This is driving some manufacturers to consider more environmentally conscious material selections, the use of more energy-efficient manufacturing processes, and the introduction of "circular" practices in design and manufacture (see Chapter 17 of ref. [10]).

Within this chapter we will explore different types of manufacturing (subtractive, formative, and additive) through case studies. We will start with traditional manufacturing methods and their place within the contemporary product landscape. We will then consider new and emerging technologies and how they can drive innovation within sporting goods design and development. The role of more contemporary technologies, such as additive manufacturing, and how they can be used for rapid prototyping or to produce bespoke items for elite athletes, will be considered (see Chapter 16 of ref. [10]). How such tools support novel design concepts, such as organically shaped and "lightweight" structures, which would be challenging or impossible to make with traditional methods, will also be discussed. The chapter considers the decisions behind selecting manufacturing techniques for a given product and concludes by summarising emerging topics in sustainable practice and "smart" manufacturing.

7.1 Subtractive manufacturing methods

The traditional manufacturing of sporting goods typically involves the process of **subtractive manufacturing**. The general concept and processes behind these methods is largely

DOI: 10.4324/9781003411086-7

detailed in existing literature [169, 170], and thus examples are presented below to exhibit key case studies, including how traditional manufacturing has developed.

Box 7.1

Subtractive manufacturing is a process in which a final product is created by removing material from a solid piece of "stock" material. These methods involve various tools and techniques to selectively remove material until the desired shape, size, and features are achieved. Broadly, the *means* of material removal is cutting, carving, milling, or turning. This is typically achieved through processes such as machining, drilling, grinding, or electrical discharge machining (EDM).

Sporting goods differ from many consumer goods due to sporting rules and regulations. These are often used to retain the history of the sport or to ensure equitable equipment between participants, or both. Such regulations might include geometric or performance attributes, which can narrow material choices for sports engineers and hence influence manufacturing. The International Cricket Council (ICC), for example, requires bats to be constructed primarily of "cane or wood". They also provide manufacturers with a "Bat Gauge" (see Figure 7.1b), which the finished bat must be able to pass through, hence

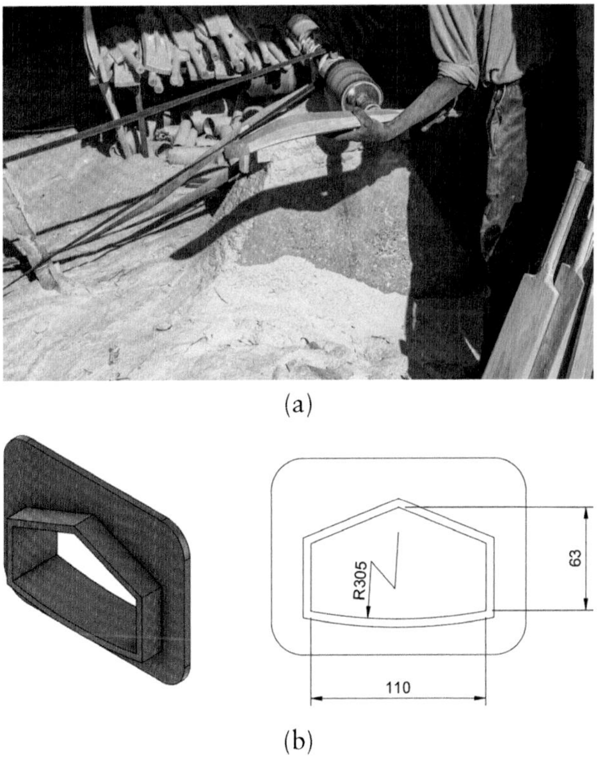

Figure 7.1 (a) Many manual techniques can be employed in shaping the cleft for a cricket bat. Appendix B.8 of the MCC laws prescribe dimensions for a bat gauge through which the blade of the bat must be able to pass (b).

limiting the size and shape of its cross section. The material and geometric specifications are set by the Marylebone Cricket Club, in Law 5 – "The Bat" [171].

Cricket bat production remains a largely human-based and artisan process (see Figure 7.1a and b), with manual techniques widely employed to shape the cleft (a solid piece of "stock" wood that makes up the bulk of the blade of the bat). In accordance with the rules, the cleft is timber (a natural material that is both anisotropic and heterogenous) and hence care is taken in the cultivation of the timbers used. Typically, willow is used as it offers the right balance of hardness, density, and moisture resistance. Manual techniques are typically used to shape the profile (within the bounds of the Bat Gauge) as well as the transition of the profile along the length of the blade [172]. Tuning these factors alters the mass distribution, impacting upon the bat's playing characteristics, including the perception of "swing weight" through the moment of inertia [173] and the impact efficiency as determined through the coefficient of restitution [174]. Some brands now use computer numerical control (CNC) techniques to shape bats [172].

7.2 Computer numerical control manufacturing

While contemporary manufacturing remains influenced by traditional methods, the modern emphasis for most sports (and particularly those manufactured in high volumes) is on efficiency in manufacture and part conformity. Within subtractive manufacturing, CNC (Figure 7.2) adds the repeatability and accuracy required and pushes these artisan methods into more precision-oriented outcomes. Here, high-quality machinery and digitised control allow manufacturers to achieve accurate and intricate shapes, ensuring tight tolerances and quality finishes for their sports products.

Figure 7.2 CNC machining a solid piece of stock material produces a lot of heat, and lots of coolant is required to prevent excessive tool wear

There are countless examples of machined products within sports. Wholly CNC machined and finished equipment often commands a higher price tag than forged (see moulding) and machine-finished alternatives. For example, some golf iron manufacturers elect to CNC machine (Figure 7.2) their clubs from a piece of forged steel stock. Each club requires several hours of machine time to complete, offering a high-precision outcome and a price tag to match.

7.3 Moulding, casting, and forming

The nature of sport means that almost all equipment sees direct human interaction. The human–equipment interface means that ergonomically designed, organic forms are prevalent throughout the industry. As a result, designs often require complex shaping with multi-directional curvature used to provide appropriate fit. Elsewhere, sports engineers will work to optimise performance, shaping products to reduce aerodynamic drag, to minimise mass, or to optimise mass distribution. For each of these, manufacturing using moulding and forming techniques is indispensable.

Moulding, casting, and forming processes typically present the highest level of consistency and reproducibility available through batch manufacturing. This ensures that each product consistently meets the required conformity standards. These techniques are typically the most cost-effective means to accommodate mass production. Injection moulding, for example, is instrumental in meeting the demands of the typically lower cost, higher production volume sporting goods consumer market. The numerous examples of injection moulding in sporting goods include elements of protective equipment, footwear components, and handles, grips, and buckles.

Box 7.2

Moulding, **casting**, and **forming** are process types that are used to shape raw materials into desired forms. These processes produce various products from high-volume, mass-produced plastic parts to individual and highly specialised composite constructions.

Moulding involves the shaping of a molten or semi-molten material using a mould. The mould serves as a template, allowing the introduced material to take on this shape as it solidifies or cures.

- **Injection moulding** is commonly used for plastics. This process involves injecting molten material into a mould, which then cools and solidifies to take its shape.
- **Compression moulding** involves placing a material, often a rubber or thermosetting plastic, into an open mould cavity and then compressing it to shape it.
- **Blow moulding** can be used for creating hollow objects, such as plastic bottles. The process involves inflating the heated plastic inside a mould to take its shape.

Casting involves pouring a molten material into a mould, where it solidifies as it cools to take its shape. Casting and moulding are similar. Their processing options are often comparable, with minor changes to suit material choice, scale, or geometric requirement.

Forming is the process of shaping solid materials without material removal. Forming typically involves forcing or bending stock materials over solid forms, deforming it into desired shapes.

- **Sheet metal forming** includes processes like bending, stamping, and deep drawing to shape sheet metal into desired forms.
- **Forging** is the shaping of metal through compression. Hot forging is done at elevated temperatures. Cold forging is typically done at or near room temperature.
- **Rolling** can be used to reduce the thickness or change the cross-section of a material by passing it through a set of rollers.

Extrusion forming is when semi-molten material is pushed or pulled through a shaped die to create continuous, uniform shapes with a constant cross-section.

The manufacturing of plastic kayaks offers an insight into why different types of moulding or forming, or both, might be selected to create a product. This industry sees rotationally moulded, blow moulded, formed, and even inflatable kayaks. Each outcome has benefits and limitations and is chosen to meet requirements in product performance and price point.

Most plastic kayaks (Figure 7.3) are produced through rotational moulding, typically using polyethylene (PE) as a material. Rotational moulding (rotomoulding) is a manufacturing technique that creates hollow parts within a rotated mould. A mould is created into which material powder or pellets are introduced. The mould is then rotated (typically across all axes using gyroscopic motion) and heated to melt the material. The rotation coats the inner surface of the mould with the molten material. The mould is slowly reduced in temperature while rotated, allowing the material to solidify and form a hollow shell. The process is cost-effective, with lower-priced moulds and less material waste than other moulding methods. Polyethylene is a durable, impact-resistant material, which makes this method popular for whitewater kayaks. Kayaks created this way are typically more flexible and heavier than those created from composite materials, and tend to be lower in cost.

Paddlers looking for stiffer, lighter plastic kayaks might choose thermoformed models. These are typically more expensive than their rotomoulded counterparts. Thermoforming uses a vacuum to draw heated sheets of material down upon moulds. Kayaks are typically moulded in two parts (deck and hull), which allows fine-tuning of material thickness for each of these elements. The individual parts are also easier to release from their moulds, permitting finer detail in the shapes created, and using less material than rotomoulding. The moulded parts are typically bonded, and the seams taped, to create a watertight kayak.

7.4 Composite manufacturing

Composite materials are commonly used in the world of high-performance sport. The benefits of creating sporting goods using these materials was outlined in Chapter 5 and Chapter 6. As discussed, composite materials offer several performance benefits, including design freedom that stems from various material selections, volume fractions, fibre orientations, and fibre placements. Importantly, the performance advantages on offer

Figure 7.3 One option with rotomoulding is to add material with a different pigment to the mould during the moulding process. This adds a (somewhat) controllable multi-colour outcome. If you ever see a multi-colour kayak like these, it is most likely rotomoulded!

differ depending on whether short or long fibre composites are used. **Short fibre composites** typically provide easier processing and more isotropic properties but generally have lower strength. Meanwhile, long fibre composites provide greater load transfer and directional stiffness, which makes them the preferred choice for many high-performance sporting applications.

Box 7.3

Short fibre composites are materials that are reinforced by adding short (typically only a few millimetres in length) reinforcing fibres (such as glass, nylon, or polyester) to a polymer matrix. These short fibres are discontinuous throughout the material and become randomly oriented during production. Short fibre composites are typically weaker than their long fibre counterparts, as the reinforcing fibres used are too short to fully transfer stresses from the matrix. The benefits of short fibre composites include their relative ease of processing (they can be produced through high-volume techniques such as injection moulding), the design flexibility to manufacture complex shapes, and cost-effectiveness over traditional materials that would provide similar mechanical properties.

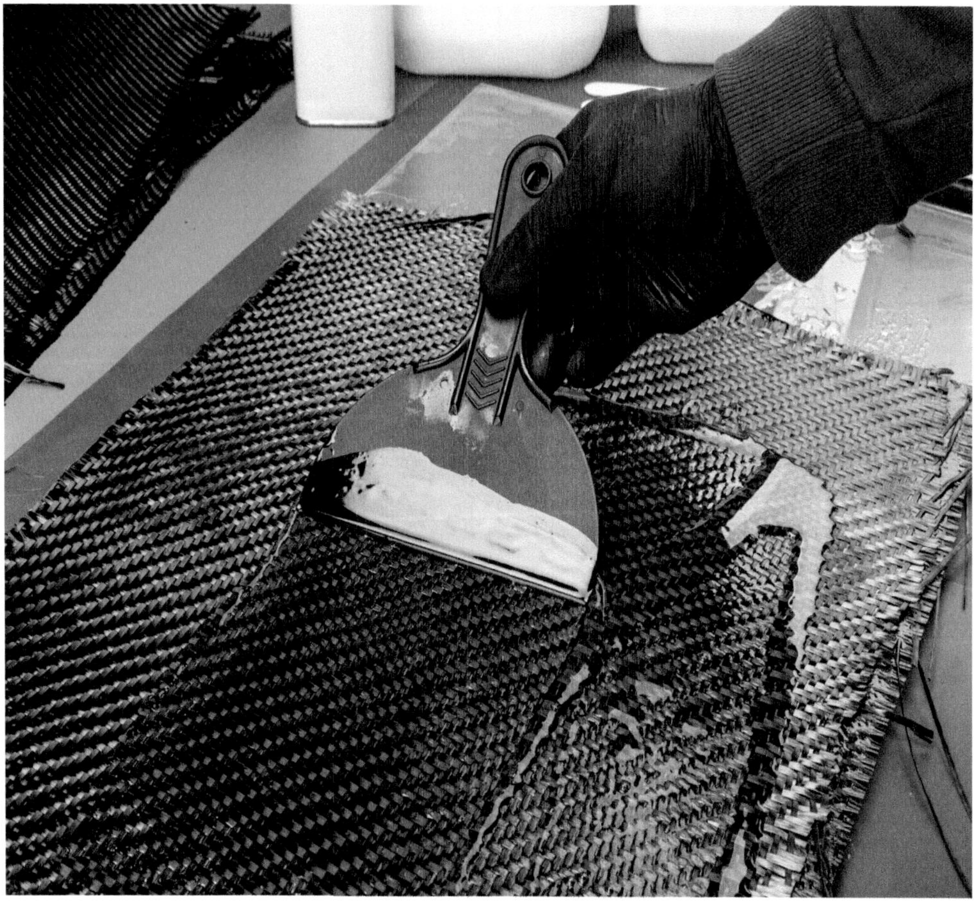

(a)

Figure 7.4 (a) Wetting dry fibres with resin during hand lay-up. In (b) we see the multi-directional curvature and clean surface finish available through compression moulding of composite parts.

(b)

Figure 7.4 (Continued)

The process of manufacturing sporting goods with long fibre fibre-matrix composites involves layers of reinforcing fibres (such as carbon fibre or fiberglass) combined with a matrix material (most commonly thermoset resins such as epoxy, polyester, or vinyl ester-based resin) (Figure 7.4a). The orientation and arrangement of fibres during manufacturing contribute to the specific characteristics of the final product. As a result, care is taken in weave selection, and during the lay-up phase (the alignment of fibre layers) to ensure the final composite has the properties required. Processing techniques including **moulding** and **filament winding** allow manufacturers to produce parts of varying complexity.

Today, composite materials are ubiquitous in high-performance sports (see Chapters 11 and 12 of ref. [10]). Much of the recent advancements in sporting goods performance owe some of their success to the introduction of composite materials in lieu of plastic, metals, or wood. From sailing masts to running blades, the ability to fine-tune material performance and reduce mass has presented clear opportunities for equipment improvement. This remains a developing space as the industry continues to explore innovative ways to leverage composite design and manufacture, e.g., the use of auxetic composites in tennis and padel rackets.

Whilst composite materials have many performance benefits, their negative impacts are also becoming better understood. Most composites use thermosetting polymer matrices, which are non-recyclable. As such, the industry is exploring more recyclable thermoplastics or the implementation of green (bio-based) composite materials, or both [15]. A key challenge in recycling fibre-matrix composite materials is separating the fibres from the matrix. If recovered, such fibres can, for example, be chopped up and used in short fibre composites.

Box 7.4

Fibre-matrix composite manufacturing for long fibre composites typically follows a moulding process where the constituent materials, usually consisting of a reinforcing material embedded within a matrix, are brought together to consolidate. Moulding techniques should ensure proper alignment, consolidation (removal of any air pockets), and curing of these components to create strong, lightweight, and durable structures. Composite manufacturing techniques include the following:

Hand lay-up is a simple, manual process where reinforcing fibres are placed into a mould, and resin is applied with brushes or rollers. The layers are consolidated manually and left to cure. This technique is often used for large, low-cost parts like boat hulls (Figure 7.4a).

Spray up simultaneously sprays chopped fibres and resin onto a mould using a spray gun. This is a quick and low-cost method that is suitable for creating large, less structurally demanding short fibre composite parts. It is uncommon within sporting goods design, other than for making low-cost boats.

In **vacuum bagging**, resin is applied to fibre weave layers and a vacuum is drawn within a sealed bag to compress the layers for better consolidation. This process improves the fibre-to-resin volume ratio and reduces voids compared to hand lay-up and spray up.

Resin transfer moulding (RTM) involves placing dry reinforcing fibres in a closed mould and injecting resin under pressure to fill the cavity. Using a closed mould ensures consistent resin flow and fibre alignment under controlled conditions. This process minimizes excess resin and air pockets, and produces parts with uniform thickness, minimal variability, and high-quality surface finish.

Filament winding is a process where continuous fibres are impregnated with resin and wound onto a rotating mandrel in specific patterns. This method creates strong, lightweight cylindrical structures and is used to make sports equipment like high-performance bicycle rims and golf club shafts.

Pultrusion is a continuous manufacturing process where fibres are pulled through a resin bath for impregnation and then shaped and cured in a heated die. This method is ideal for producing profiles like rods, beams, and sporting goods with consistent cross-sections.

Finally, with **compression moulding,** pre-impregnated (fibres already impregnated with matrix material and partially cured) are arranged within a mould structure and heated with pressure applied to shape and cure the part. This technique is particularly efficient for mass producing parts with complex shapes and is commonly used within the sporting goods industry (Figure 7.4b).

7.5 Additive manufacturing

Additive manufacturing (colloquially referred to as 3D printing) encompasses various manufacturing technologies. These typically deposit or consolidate material in layers to build up a 3D part in thin slices. Additive manufacturing technologies are versatile and can produce intricate geometries that might be impossible to achieve with traditional manufacturing techniques. Furthermore, the internal architecture of the part can be designed, allowing the creation of complex parts that are customised for specific users or applications. Additive manufacturing offers several benefits to sports engineers, offering complex geometries, quick and cost-effective design reiteration, on-demand and localised manufacturing, and part customisation for limited additional cost. These benefits have made additive manufacturing a popular topic within the sporting goods design community [175]. Despite this, additive manufacturing has disadvantages [176], many of which have limited the roll-out of larger-scale production of consumer products. Table 7.1 compares additive and subtractive manufacturing across various benchmarks, providing a snapshot of the capabilities available to sports engineers through each technology.

Novel additive manufacturing technologies are continually being developed to suit specific applications. Today, the scale and application of additive manufacturing are broad, from desktop polymer printers available for purchase at high street retailers to multi-metal printing facilities [177], and specialised techniques for complex metal repair [178]. Through this landscape, three processes are the most prominent in terms of industrial use, versatility, and application: material extrusion, vat photopolymerisation, and Powder Bed Fusion (PBF).

Table 7.1 A comparison of additive and traditional (subtractive) manufacturing methods

Aspect	*Additive manufacturing*	*Traditional manufacturing*
Process	Layer-by-layer deposition using processes like fused filament fabrication, stereolithography, digital light processing, and powder bed fusion, guided by a computer-aided design (CAD) model.	Subtractive machining (milling, turning, grinding) or formative techniques (casting, forging). For CNC outcomes this will also be guided by a CAD model.
Design complexity	Near-unlimited design freedom; enables lattice structures, internal channels, and lightweight optimisation.	Design may be constrained by tool paths and tool access (machining), or by mould geometry – which is typically created through machining.
Material utilisation	Typically >95% material efficiency; minimal waste.	Material utilisation varies, often <50% for subtractive processes.
Production scalability	Low scalability; suited to single items, small batch runs, or iterative prototyping.	Highly scalable; more efficient for high-volume production (particularly in moulding and forming processes).
Setup requirements	Requires CAD file preparation, slicing software, and machine setup.	Requirements vary, but may include mould/die design, CNC programming, and material preparation.

(Continued)

Table 7.1 (Continued)

Aspect	*Additive manufacturing*	*Traditional manufacturing*
Cost dynamics	Low tooling costs; high material costs, particularly for highly processed PBF powders.	High initial tooling costs (particularly for moulded parts), which may be recovered over large production volumes.
Dimensional accuracy	Typically ±0.1–0.2 mm, depending on the manufacturing process and material.	High precision, often ±0.01 mm with CNC machining.
Surface finish	Rough surfaces (Ra 10~50 μm) due to layer stepping; typically requires post-processing for smoothness.	Superior as-machined finish (Ra 0.4~3.2 μm) or near-net shape in casting.
Mechanical properties	Anisotropic properties; strength typically depends on layer adhesion, material type, and printer settings.	Isotropic properties in machined or forged parts; predictable material performance.
Lead time	Rapid for initial parts; dependent on build volume and layer resolution.	Rapid for machining. Can be lengthy to create tooling for moulding but faster for bulk production once setup is complete.
Customisation	Suited to bespoke components, prototyping, and unique geometries.	Machining is also well suited to bespoke components and prototyping. Larger volume moulding methods offer limited customisation and require significant retooling for design changes.
Environmental impact	Lower material waste but higher energy demand for processes like PBF.	Higher waste due to scrap material; potentially less energy-intensive for bulk production.

7.5.1 *Material extrusion*

The most common additive technology is based upon the principle of melting of thermoplastic material and extruding it through a nozzle to create a material path in layers. This is often referred to as Fused Filament Fabrication, as stock material is most typically created as a long strand of filament wound upon a spool. The technology (Figure 7.5a) is common for creating early prototypes [179] and models (Figure 7.5b), including those used in sporting goods design studies [180]. Unfortunately, it often suffers from high material anisotropy and (relatively) poor surface quality. Despite promising outcomes in some tests [181], material extrusion is rarely used commercially to manufacture sporting goods.

(a)

(b)

Figure 7.5 (a) Material extrusion can create complex shapes and accommodate overhang, provided the overhang angle is not too high. (b) A wide range of thermoplastic materials (and colours!) can be used.

7.5.2 *Vat photopolymerisation*

Photopolymerisation is the oldest additive manufacturing technology. It uses a directed energy source to solidify a specially selected photosensitive polymer and build a 3D object. There are two main classes of equipment within this technology space, Stereolithography

(SLA), and Digital Light Processing (DLP). In SLA, an ultraviolet laser is "drawn" across each layer of a print using articulated mirrors. This solidifies the liquid polymer and hence builds a layer of the 3D print. The build platform then moves one increment in the height direction, and the process is repeated with the geometry for the next layer. The part slowly gains mass during the printing process, so support structures are required to secure it. These structures are created from the same photosensitive material, which means their removal after printing may leave "witness" marks on the surface of an otherwise high-quality, finished part.

Digital Light Processing is similar to SLA but, instead of using a single point laser, uses a light projector to project an entire layer of a print at a time. This means that DLP can complete each slice quickly and hence is typically faster at creating parts than the equivalent SLA printer. In either technology, printed parts require cleaning and final curing to consolidate them. Once complete, the thermoset polymers used are typically more stable, and the parts closer to being isotopic than those created through material extrusion. Flexible resins used in vat polymerisation allow for the creation of rubber-like parts with good elongation and impact resistance, making them ideal for soft-touch components (Figure 7.6a) or functional seals.

7.5.3 Powder Bed Fusion

Powder Bed Fusion encapsulates a collection of technologies that revolve around the melting or sintering of powder particles to create a solid object. Typically, a laser or electron beam is passed over a layer of powdered material, solidifying it along the path. Further layers are created over the first, creating a solid volume of material, and describing a 3D object. As each consolidated layer is created from a "full" layer of powder, parts are self-supporting, which (for polymers in particular) often mitigates the need for support structures and allows for complex designs. Multiple parts may also be "nested" above one another within the build volume, which can increase the efficiency of the printing process.

Some forms of PBF allow printing of metal parts with complex geometries. These typically require support structures to prevent them from distorting due to temperature gradients across the powder bed. Various metals, including steel (Figure 7.6b), stainless steel, titanium, and aluminium, can be printed. This means that PBF is the most common additive manufacturing technology used for creating structural parts.

7.5.4 The impact of additive manufacturing on sport

The potential benefits of additive manufacturing have been documented for some time, yet its commercial uptake for full-scale manufacturing remains limited [182]. Despite this, many new additively manufactured sport products are brought to market each year, and the bulk of published research remains positive about the potential benefits of successfully implementing these techniques. Novak and Novak [175] proposed that research regarding widespread implementation of additive manufacturing within sports is less well developed than within other fields (such as medical device design). They highlighted several opportunities and challenges for the future of additive manufacturing in sports (Table 7.2).

(a) (b)

Figure 7.6 Vat photopolymerisation can be used to create pliable polymeric structures such as (a) the bicycle seat shown here. Meanwhile Powder Bed Fusion can be used with metals to create products such as (b) golf club heads, where material hardness is critical to product longevity.

Table 7.2 Opportunities and challenges of additive manufacturing of sporting goods adapted from Novak and Novak [175]

Opportunities	*Challenges*
Creation of athlete-specific geometries	Limited data on long-term material durability
Digital design and creation allow for simple geometry updates	Typically, a slower end-to-end process than traditional manufacturing
Part consolidation can reduce assembly costs	Post-processing is often slow and labour-intensive
Ability to create complex and novel geometries to improve performance	Additive manufactured parts often interface with traditionally made parts, reducing design freedom
	"Automated" optimisation methods are often more manually driven than advertised
	New products may fail to meet traditional sports regulations
	Laboratory testing of proposed products may not replicate real-world conditions

7.6 Case study: Additive manufacturing in sport

This case study considers the failure of an additively manufactured bike component. It provides insight into why the technology was chosen and to the potential benefits, as well as serving as a reminder of some of the inherent dangers when quickly bringing prototype parts into sporting use. During qualifying for the team pursuit track cycling event at the 2021 Tokyo Olympics, a set of additively manufactured handlebars experienced a sudden failure. This caused a crash for Australian cyclist Alex Porter. Following the incident AusCycling, the Australian governing body for cycling, published a report exploring the mechanisms behind this high-profile failure [183].

The report explained how the four-person pursuit team had recently switched to using smaller bicycle frames to gain an aerodynamic advantage. This change unfortunately led to the fourth rider's knees contacting the handlebar during the high torque standing start phase of the event. A proposal was made to lengthen the base handlebar to provide more clearance. The short timeframe and uncertainty of timely delivery from the original supplier of the moulded handlebar led them to pursue a locally (Australian) made additively manufactured replacement. The revised part was printed in Ti-6Al4V titanium using PBF, to an outline design and specifications set by AusCycling that were *broadly* (see key inconsistencies below) in line with the relevant standard for bicycle handlebars (ISO 4210).

The investigation found no problems with the design or manufacture of the handlebar, but highlighted several inconsistencies in process that contributed to the failure:

- The loads stipulated in ISO 4210 were set based on the use of traditional "drop" handlebars. The aerodynamic shape of the handlebar that failed, combined with the elite athletes, meant it experienced higher loads (close to 150%) during use.
- The fatigue testing requirement of the new handlebar was reduced from the 200,000 cycles stipulated in ISO 4210 to 50,000 cycles to expedite delivery.
- The supplier of the bar provided no information about its mounting torque, and the value used was likely too low for secure attachment to the existing stem structure.

This case study shows the role of additive manufacturing in providing bespoke, customisable, and short-lead-time sports products. It also highlights the pitfalls of human expectation, and the need for a systems engineering approach to sporting goods design (i.e., could the rider size issue with the smaller bicycle frame have been foreseen?). As modern manufacturing shortens the lead time on sports products, we risk expecting high-performance outcomes in unrealistic timeframes. Paradoxically, the ability to design custom geometry means that more time should be spent in understanding the specific use case(s) of the sports product. This means that exploring performance through design, simulation, and product testing is increasingly important. Designing for additive manufacturing is inherently different to that for traditional methods [184].

7.7 Environmental sustainability in sporting goods manufacturing

From raw material extraction to end-of-life disposal, each stage of the sporting goods lifecycle has an inherent environmental impact [15]. As we learn more about the impact of our actions on climate change, we have started to understand how our choices can influence the occurrence of natural disasters in our changing world. Increasingly, many sporting goods manufacturers are recognising their role and understanding the substantial environmental (and possible financial) benefits available through more sustainable business practices. Over time, more sporting goods manufacturers are engaging in sustainable manufacturing. Sporting goods is a wide-ranging remit with a wide consumer base. The industry has examples of companies that pursue ultra high performance (often at the expense of environmental impact) as well as ultra low environmental impact companies and sports products (which may come at the expense of maximal performance). It should be stated that the bulk of sporting goods production sits at neither end of this spectrum, and the highest volume of production for sporting goods is sportswear, anoraks, and gymnastic equipment [185], typically produced in Asia, Eastern Europe, and emerging countries (such as Vietnam and Indonesia).

Consumer awareness is one of the driving forces behind growth in sustainable practice. The so-called "Blue Planet Effect" saw an increase in attention on sustainable practice following clear documentation of human impact on our environment. Whilst specific consumer behaviours have been harder to determine [186], the general trend towards awareness has incentivised some sporting goods manufacturers to encapsulate sustainable practice amongst their values. Unfortunately, pressure for rapid innovation within sports typically creates products with short lifecycles. Similarly, marketing practices and material innovation in high-performance sporting goods can also create barriers to recycling and reuse [187].

> Box 7.5
>
> **Sustainable manufacturing** is a process through which the creation of products is undertaken with the environment in mind. For example, the United States' Environmental Protection Agency (EPA) stipulates that sustainable manufacturing should encompass the minimisation of negative environmental impact, through the conservation of energy and natural resources. Sustainable manufacturing should also encapsulate human aspects of sustainability, such as product safety, employee, and community support.

Materials and their reuse are areas of particular interest to sporting goods manufacturers and consumers. Unfortunately, the specialised materials commonly used in sporting goods are often designed primarily for high performance, durability, or safety, which can present challenges in reuse. In addition, parts are often made from composites, bonded materials, or tailored synthetics that are not easily separable at end of life [15]. Supply

> Box 7.6
>
> **Drivers and incentives** are factors or motivations that encourage or influence behaviour, decisions, and actions. Today, we see more companies objectivising sustainable practice, which is likely driven by a mix of regulatory, economic, reputational, and market-based incentives. The relative importance of each driver depends on factors like market size, company visibility, and production processes.
>
> **Regulatory pressures** are particularly relevant for mass-produced goods like sportswear, footwear, and packaging, where recycled content laws and producer responsibility schemes are influencing design.
>
> **Economic factors** (such as cost saving and resource security) are important in sporting goods manufacturing, particularly where materials like aluminium, rubber, or plastic are used at scale. Recycled materials can be more affordable when virgin material prices are high, and building supply chains around recycled inputs can protect manufacturers against resource scarcity and price volatility.

Reputational incentives stemming from heightened consumer awareness of sustainable practice are a behaviour driver, particularly for lifestyle and outdoor gear. Many consumers now *expect* sporting goods brands to prioritise sustainability, and they face increasing pressure from investors and the public to show environmental leadership. Leadership might include the adoption of *voluntary* standards and certifications (e.g., ecolabels [15, 188]), which have grown in prevalence, particularly in areas like textiles and footwear, where long global supply chains require increasing accountability.

Market-based incentives linked to innovation and competitive advantage can also be aligned with sustainable practice, where companies might explore new materials and design approaches to offer sustainable products without compromising performance.

chain limitations and cost pressures also often make it difficult for smaller manufacturers to access or adopt circular solutions. While reusing materials in sporting goods can be challenging, it may be feasible in specific applications, e.g., where waste streams are consistent, or performance requirements can still be met.

The construction of athletics surfaces from recycled rubber crumb is an example of waste material reuse. The rubber is processed by shredding automotive tyres (Figure 7.7a) into small granules, usually 1–4 mm in size (Figure 7.7b), with steel and textile fibres removed during this process. These clean rubber particles are then mixed with a polyurethane binder to form a durable, weather-resistant material (Figure 7.7c). In athletic tracks, recycled rubber crumb is commonly used as the base or cushioning layer. In more advanced systems, the rubber crumb base may be overlaid with virgin materials to tune performance requirements for traction, ultraviolet light resistance, or overall durability. Rubber crumb is also used as a performance infill in artificial turf surfaces. However, new EU regulations restricting intentionally added microplastics (Commission Regulation (EU) 2023/2055) mean sports engineers are now seeking more environmentally friendly alternatives [189–191]. Another prominent example is the use of recycled polyester (which is typically made from post-consumer plastic bottles) in sports apparel. Major brands have incorporated this material as common practice, while others are exploring products created from waste plastics collected from coastal environments [192].

Recycling of products constructed from high-performance materials remains a challenge. Fibre-matrix composites are used widely within sports equipment, to the extent that the sports equipment industry is the fourth largest consumer of carbon fibre-reinforced polymers (behind aerospace, automotive, and wind energy). In 2018, sports equipment was responsible for 13% of such composite usage worldwide [193]. Recycling these composites is possible but is labour- and energy-intensive. Thermoset composites are typically recycled by mechanically shredding or milling them into small flakes or powders. This produces a material that can be used as a filler for injection moulding, but as a short fibre composite with somewhat degraded mechanical properties (i.e., downcycling). More resource-intensive methods such as heating via pyrolysis

(a)

(b)

(c)

(d)

Figure 7.7 Discarded automotive tyres (a) are granulated (b) before being combined with a binder agent and laid for sports surfaces (c). Discarded plastic waste is spun into fibres (d) for use in sports garments.

(in the absence of oxygen) to break down the resin, heating via hot gas to burn off the resin, or dissolving the resin component in solvents at high temperature may recover longer fibres with better properties, but at the expense of cost. In principle, thermoplastic composites can be recycled multiple times since their matrix re-melts. In practice, even thermoplastic composites are predominantly recycled via mechanical means, turning the composite into flakes or pellets. These can be directly re-melted and re-processed for moulding, but once again, the fibre length substantially shortens during processing, degrading the properties considerably compared to the continuous-fibre original products.

Box 7.7

Greenwashing is when a company or product is marketed as more environmentally friendly than it really is, often by exaggerating or selectively highlighting "green" qualities. Greenwashing creates the impression of sustainability without substantial actions to back it up, misleading consumers and distracting from genuine environmental solutions.

Lyon and Maxwell [194] describe greenwashing as "the selective disclosure of positive information about a company's environmental or social performance, without fully disclosing the negative information on these dimensions, so as to create an overly positive corporate image".

The combination of improved recyclability and reduction in waste during manufacture has led some sporting goods manufacturers to introduce thermoplastic composite products on the basis of their green credentials when compared to their thermoset counterparts. Sports engineers should be mindful to evaluate these claims carefully. This is because composite products marketed to reduce consumer guilt [195] thanks to their environmental benefits may be at risk of **greenwashing**, as composite re-processing for re- or upcycling is challenging [196], and hence re-processing is more often a process of **downcycling**.

Box 7.8

Downcycling refers to the recycling of materials into products of lower quality, reduced functionality, or diminished economic value compared to the original. Downcycling typically results in degraded mechanical, chemical, or aesthetic characteristics that limit future uses. Whilst this process helps divert waste from landfills, creating "greener" products that can ultimately only be downcycled does little to eliminate overall resource depletion.

Adopting sustainable practices in sporting goods manufacturing remains a continuing challenge, despite growing awareness and effort across the industry. Amid these challenges, recycled polyester stands out as one of the most impactful and scalable solutions currently available, not least because plastic waste can be spun into fibres (Figure 7.7d) with similar properties to virgin material. Its wide adoption and alignment with existing manufacturing infrastructure gives recycled polyester a role as an exemplar material. This can help the wider sporting goods industry make meaningful progress even at lower price points while longer-term recycling technologies continue to develop.

7.8 Emerging topics and future manufacturing trends

"Advanced" manufacturing is a broad term that refers to the use of cutting-edge technologies and innovative processes to enhance efficiency and precision in manufacturing processes. What formally constitutes advanced manufacturing remains loosely defined. Broadly, these are techniques that expand upon traditional methods, incorporating modern technologies to improve product quality and reduce manufacturing costs through efficiency savings (speed of manufacturing processes) and reduced waste (both process waste and through tighter quality conformity control leading to fewer rejected parts).

As technologies mature and become more widely adopted, the boundary between traditional and advanced methods shifts. Digitalisation and "Industry 4.0" have driven much of this recent development. Here, the use of connected sensors and data-driven systems enable "smart" factories in which machines monitor their own wear, select tools, and even manage stock inventory in real time (e.g., Figure 7.8). Automation and robotics play a key role in these factories, particularly in repetitive and high-precision tasks. The improved consistency, conformity, and throughput of such factories can also support broader sustainability goals, including energy efficiency and waste reduction.

While these technologies have seen widespread uptake in certain sectors (e.g., automotive and consumer electronics), their uptake within sporting goods remains limited. There are some exceptions (e.g., sports wearables and electronic devices, although these are arguably a subset of consumer electronics), but much of the manufacture of conventional sporting goods continues to rely on traditional manufacturing methods. The time-intensive and manual nature of these methods means that sporting goods manufacture is typically undertaken in global regions with lower labour costs.

The future of any manufacturing segment can be hard to predict, but recent supply chain disruptions (such as those caused by the Covid-19 pandemic, global shipping delays, and changes in political regimes) have highlighted the vulnerabilities of offshore manufacturing and increased demand for more responsive supply chains. Some consumers are also showing stronger preferences for locally made sporting goods. This is often motivated by a perception of higher quality or concerns about overseas sustainability and ethical labour practices, or both. It may yet be the case that the efficiencies available through smart manufacturing permit **reshoring** of sporting goods manufacture. While cost currently remains a barrier, there is increasing interest in reshoring sporting good manufacture, particularly for premium or custom items.

Figure 7.8 Industry 4.0 revolves around the digitalisation of manufacturing spaces. Digital technologies are integrated into the manufacture workpiece to monitor process and optimise outcomes in real time.

Box 7.9

Many sporting goods companies started in the Western world and, over time, migrated manufacturing overseas, to regions with lower-cost labour. **Reshoring** is the process of bringing these activities back to a company's home country. It is often driven by the desire to reduce supply chain risks, improve product quality (whether actual or perceived), shorten lead times, and respond to increasing consumer demand for locally made, ethically produced goods.

7.9 Manufacturing trends: A case study in lugged bicycle construction

"Lugged" construction has a long history in bicycle manufacturing, tracing back to the turn of the 20th century when steel was the predominant material for frames (Figure 7.9a). Traditionally, lugs were cast or stamped metal sleeves that allowed frame builders to join steel tubes using brazing – a process where a filler material, but not the tubes, is melted to bond the components. This technique provided strong, precisely aligned, durable joints. Over time, improvements in gas tungsten arc welding (TIG) and post-welding heat treatments led to rapid uptake of aluminium alloy framesets. Metal-forming technologies also improved, allowing manufacturers to move away from regular tubes (plain hollow cylinders) and into more structurally efficient forms with varying cross-sections. Advances in fibre-matrix composite moulding accelerated this change, allowing manufacturers to form complex frame geometries (Figure 7.9b).

In the mid-2010s, the growing capabilities of additive manufacturing saw some bike brands exploring the technology. The first widely known example is Empire's MX-6 Evo, created with manufacturing partner Renishaw [197]. The titanium frame was created via PBF within a build volume of 300 × 250 × 250 mm. Accordingly, it was built in pieces, then bonded together. Renishaw listed numerous advantages of this process, including [197]:

- Rapid iterations, giving flexibility to make design improvements right up to production
- Ability to make shapes derived by topological optimisation
- Ultimate customisation and tailoring, allowing one-offs to be made as easily as production batches

Fully additively manufactured frames failed to gain uptake in the bicycle industry, but the advantages remained attractive. Accordingly, lugged construction regained some popularity, with additive manufacturing used to create lugged geometries (Figure 7.10)

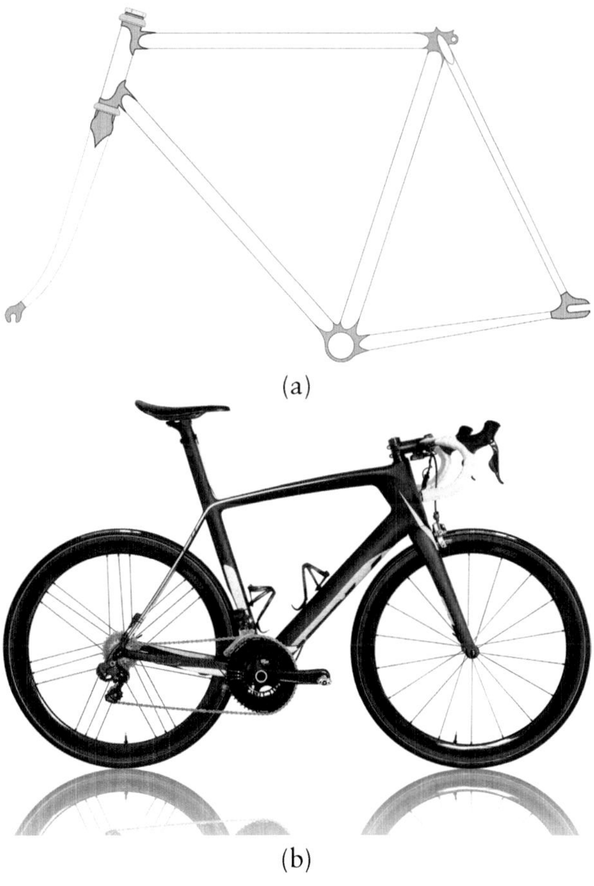

Figure 7.9 (a) A drawing of a road bike highlights the lugged construction methods that were traditionally used to produce (typically steel) bicycle frames. In (b) we see the silhouette of a newer frame, highlighting the radical changes in geometry available through moulding of fibre-matrix composites.

Figure 7.10 Additive technologies such as Powder Bed Fusion (PBF) allow manufacturers to create customised lug geometries that can create an athlete-specific outcome through geometric and structural optimisation.

that could be tailored to individual riders or specific performance needs. Printing allows for lugs with internal lattice structures and organic forms that can help to optimise strength-to-weight ratios while integrating the aesthetic, functional, and geometric design requirements that are critical for custom bicycles. The PBF technologies harnessed advanced materials like titanium alloys and facilitated rapid prototyping and small-batch manufacture without extensive tooling, which appealed to boutique cycling brands seeking market differentiation.

Over time, the appeal of additive manufacturing spread further within the bicycle industry. This is most notable in the high-performance discipline of downhill mountain biking, where brands engage in rapid innovation and advanced prototyping to gain a competitive advantage. Prototype downhill bikes made using lugged construction methods have become increasingly common in Union Cycliste Internationale (UCI) World Cup races. Interestingly, and perhaps in response to some of the challenges of additive manufacturing (e.g., microstructural inconsistencies that may impact durability, and challenging post-processing requirements), some manufacturers have transitioned back to the more established technology of CNC machining to create prototype lugged bicycles. CNC technology offers high precision, material consistency, and typically requires less post-processing.

While advances in additive manufacturing materials and processes may eventually overcome some of the inherent limitations, this case study essentially highlights the dynamic nature of the sporting goods industry. Through exploration of a developing technology, new ways to exploit traditional technologies have been found, offering a pragmatic approach to achieving high standards of quality and performance. This development underscores a continuous balance and trade-off between innovation and practicality, as sports engineers adapt to meet the demands of rapid innovation, performance, durability, and aesthetics.

7.10 Chapter summary

This chapter provided a broad introduction to manufacturing techniques for sporting goods, beginning with traditional methods and their continued relevance in today's product landscape. It then explored new and emerging technologies that are reshaping the design and development of sporting goods. Particular attention was given to the role of additive manufacturing in supporting rapid prototyping and enabling bespoke solutions for athletes. The chapter concluded by outlining how these technologies facilitate innovative design approaches – such as lightweight, and organic structures – that were previously difficult or unfeasible using conventional manufacturing techniques. There is more to learn about the manufacturing of sporting goods than the basics covered here, some of which can be found in the suggested further reading.

7.11 Suggested further reading

The following suggestions for further reading go deeper into the topics introduced in this chapter:

- *A Practical Guide to Design for Additive Manufacturing* [198]
- *Design and Manufacture of Fibre-Reinforced Composites* [199]
- "DfAM Strategic design considerations" [184]
- *Routledge Handbook of Sports Technology and Engineering*, 1st and 2nd editions [10, 86]
- *Manufacturing and Design* [200]
- *Manufacturing Processes for Design Professionals* [169]
- *Manufacturing Engineering and Technology*, 8th edition [170]

8 Testing sporting goods

Testing of sporting goods plays a vital role in sports engineering. It provides the empirical foundation for the development, validation, and continual refinement of sporting goods. Through systematic testing, sports engineers ensure that sporting goods not only fulfil their intended purpose but also meet safety, reliability, and performance expectations. Indeed, "Test, test and test again!" was one of the guiding principles of adidas company founder Adolf "Adi" Dassler (1900–1978). Whether it is a football, a tennis racket, or a bicycle, every sporting goods item must meet specific requirements before it can be used in training or competition. These requirements can be set by sporting governing bodies, standardisation organisations, manufacturers, or the athletes themselves. Testing allows sports engineers to verify that sporting goods and facilities meet those requirements. Sporting goods testing can be broadly divided into two categories:

- Quantitative testing, which uses instruments, sensors, or simulations to generate objective measurements
- Qualitative testing, which is based on the perception and experience of athletes and users

Within quantitative testing, we can further distinguish between physical testing in laboratory settings (e.g., for stiffness, strength, impact load limitation, breathability, or water resistance), virtual testing through computer-based simulations such as Finite Element Analysis, and biomechanical testing, which examines how equipment interacts with the human body. Field testing takes place in "real-world" environments and captures performance during actual use.

This chapter provides an overview of various testing methods – ranging from controlled laboratory experiments to in-field measurements and virtual simulations. Through selected examples and applications, the role of testing in supporting the safe, effective, and rule-compliant development of sporting goods is illustrated.

8.1 Purposes for testing: Safety, compliance, and functionality

The testing of sporting goods serves several purposes. Testing ensures that equipment is safe to use, complies with relevant rules and regulations, and fulfils its intended technical functions in a sporting context. Understanding these purposes helps sports engineers to design and select appropriate test methods. In this section, we explore three core motivations for testing (see Box 8.1).

DOI: 10.4324/9781003411086-8

Box 8.1

Safety includes identifying and reducing risks associated with product use, especially under extreme or unexpected conditions. Safety testing therefore focuses on identifying equipment failure modes, understanding injury mechanisms, and quantifying protective effects.

Compliance testing is typically guided by detailed protocols defined by international standards or sporting governing body regulations. It ensures fair competition and legal acceptance. Test specifications may cover dimensions, material characteristics, performance thresholds, and durability. Non-compliance can result in disqualification of products or teams.

Functionality concerns assessing whether a product performs its intended function effectively under realistic usage scenarios. These tests often combine objective data collection with athlete feedback and may take place in laboratory, field, or virtual environments.

Each of these purposes contributes to product quality, athlete well-being, and trust in sport technology. In practice, multiple test types are often required to comprehensively evaluate a product.

8.1.1 Safety

Safety is a non-negotiable requirement when developing and approving sporting goods. One aspect of safety is structural integrity: the ability of a product to withstand loads without failure. For example, bicycle frames are tested according to ISO 4210-6:2023 [201], which includes cyclic loading in the steering head and bottom bracket areas to assess fatigue resistance. High loads, such as those occurring during landings or jumps in mountain biking, are mechanically simulated to ensure the product will not fail under extreme but realistic conditions. Comparing damage patterns from laboratory tests with real-world product failures helps validate the test method, improve the reliability of structural integrity assessments, and enhance safety.

Another critical aspect of safety is the prevention of injury during regular use or in case of accidents. Some sporting goods are designed specifically for protection – such as helmets, and body and wrist protectors [57, 202] – while others must not cause harm through poor design or degradation over time. For example, sports flooring must offer appropriate levels of stiffness to protect joints and reduce impact loads, depending on its use case (e.g., athletics tracks vs. gym floors). In some cases, safety mechanisms are included to protect third parties – such as ski brakes that prevent loose skis from sliding downhill after a fall and hitting other skiers.

8.1.2 Compliance

As well as safety, sporting goods must meet legal and regulatory requirements. These may include **national or international legislation**. Examples include the **EU Batteries Regulation (2023/1542)**, which governs battery safety, disposal, and sustainability in wearable devices, and the **EU Regulation on Machinery** (1230/2023), which must be observed when sporting goods have motors, such as e-bikes. Depending on the market, products must pass safety inspections or carry certification marks (such as CE in Europe or UKCA in the United Kingdom). Testing helps verify that these legal and technical criteria are fulfilled. Product testing ensures that such regulations are met, allowing goods to be marketed and sold legally.

Another important aspect of compliance is adherence to **sport-specific rules** set by sporting governing bodies. These rules often define technical properties such as dimensions, mass, surface characteristics, and impact behaviour. For example, World Rugby defines safety requirements for protective equipment such as headgear and body padding [203], including impact force thresholds. Similarly, the International Ski and Snowboard Federation (FIS) specify technical properties for skis used in competition [204], such as minimum sidecut radius and maximum binding height, to ensure fair and safe racing conditions. Comparable rules exist across most organised sports. Compliance testing ensures that equipment does not offer an unfair advantage and promotes a level playing field – particularly in elite and professional sport.

8.1.3 Functionality

Beyond meeting safety and legal requirements, a sports product must be functional. Functionality describes the degree to which a sports product fulfils its intended technical functions – in a similar way to how quality is defined as the degree to which a set of characteristics fulfils specified requirements. This includes contributing to performance, safety, comfort, ergonomics, or accessibility, depending on the use case and the target user.

Functionality is not a binary feature. It is measurable and context-dependent. A racing bike designed to be used in the UCI Downhill World Cup must fulfil different functional requirements to a bike designed to help a beginner learn the sport comfortably and safely. As they must do when designing a product, the sports engineer must consider the intended purpose, environment, and end-user when planning tests for functionality. Table 8.1 gives practical examples of functionality across different categories.

Testing should be aligned with the functional goals of the equipment. For instance, equipment intended to support high-performance athletes under extreme conditions must

Table 8.1 Practical examples of functionality across different categories

Functional goal	*Example equipment*	*Typical testing focus*
Performance	Sprint spikes	Energy return, stiffness
Safety	Ski helmet	Impact load limitation, strap strength
Comfort	Running shoe	Pressure distribution, breathability
Ergonomics	Rowing handle	Grip shape, wrist posture, load symmetry
Accessibility and enjoyment	Oversized tennis racket	"Forgiveness", "control", swing weight
Regulatory compliance	FIFA-approved football	Rebound height, circumference, mass

be tested differently than an item for recreational users. Only through purposeful testing can engineers ensure that sporting goods deliver value, safety, and fairness during sport.

8.1.4 Standards in context – ISO 4980:2023

The ISO 4980 standard [205] introduces a structured approach for evaluating benefits and risks associated with sports and recreational products and activities. Rather than treating safety as an absolute requirement, it proposes that risks must be balanced against functional, psychological, and social benefits. In sports engineering, this perspective can help guide testing priorities – ensuring that improvements in performance or user experience are not unintentionally offset by new risks, and vice versa. The standard encourages transparent benefit–risk documentation, especially for equipment intended for vulnerable or untrained users.

Box 8.2

Case example: Helmet development (purpose)

In developing a new high-performance cycling helmet, the engineering team defines four primary goals: (1) compliance with EN 1078:2012 [206] for certification, (2) enhanced protection against oblique impacts (not covered in EN 1078:2012), (3) a measurable reduction in aerodynamic drag, and (4) improved ventilation performance. These goals shape the testing strategy.

To achieve these objectives, the development process is structured around parallel workstreams involving material selection, simulation, prototyping, and test planning. For safety and compliance, the team studies the EN 1078 protocol and consults injury biomechanics literature to address rotational acceleration thresholds, which are not yet fully covered by standard norms. For drag reduction, engineers collaborate with aerodynamicists to define performance targets based on comparative Computational Fluid Dynamics data from competitor helmets. Ventilation is addressed both through airflow simulations and physiological feedback gathered in early concept trials.

A definition of success criteria is established: peak linear accelerations must stay below 250 g in all standard test locations; rotational accelerations are to be kept under 6,000 rad/s^2; the helmet should yield at least a 10% reduction in aerodynamic drag force at 45 km/h compared to the last season's model; and internal temperature rise under high-exertion field conditions should be at least 1.5°C lower than the previous model. These quantifiable goals inform the selection of instrumentation, test protocols, and analysis methods throughout the project.

8.2 Test methods

Various methods are available for testing sporting goods, each with their own advantages, limitations, and application domains [207]. This section presents the most important categories of test methods used by sports engineers: mechanical laboratory testing,

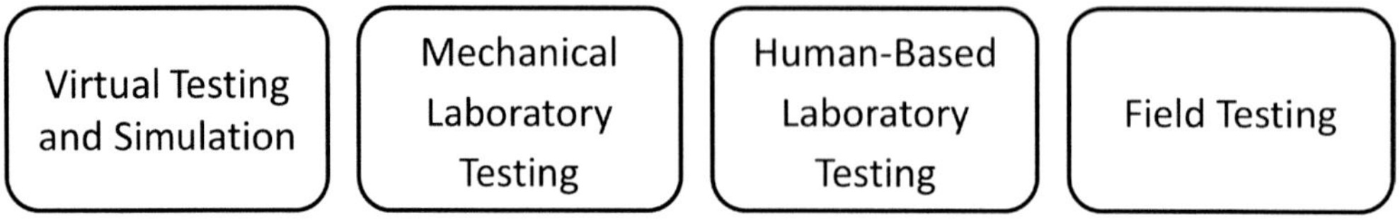

Figure 8.1 Diagram summarising four different test methods in sports engineering, (1) virtual testing and simulation, (2) mechanical laboratory testing, (3) human-based testing, and (4) field testing in sport environment.

virtual testing and simulation, human-based laboratory testing, and field testing (Figure 8.1). While laboratory methods offer controlled conditions and high repeatability, simulations allow early design refinement and insight into internal processes. Human-based and field tests, on the other hand, provide valuable information about how sporting goods perform in real use, interacting with athletes and the environment. Together, these approaches enable sports engineers to assess safety, functionality, and compliance across all stages of product development.

In modern sports engineering, virtual testing has become a powerful and increasingly common tool. Instead of just building and testing physical prototypes, sports engineers can now use computer simulations to predict how a product will behave under different conditions. This method is especially valuable in the early stages of product development, where it helps to explore different design ideas, optimise performance, and avoid costly trial-and-error processes. At this stage simulation can reduce reliance on testing prototypes, and lead to better outcomes. Two common virtual testing methods are Finite Element Analysis and Computational Fluid Dynamics (see Chapters 3, 4, and 6 of ref. [10]). Finite Element Analysis is used to understand how structures behave under mechanical loads, e.g., how a bike frame or ski flexes when loaded. Computational Fluid Dynamics is used to analyse how air or water flows around or through a product and the user, e.g., how air flows over a cyclist or skier.

Finite Element Analysis is a numerical simulation technique that breaks a complex object into many small elements, usually using a mesh of 2D or 3D shapes. These elements are connected by nodes, and physical laws – such as those governing stress, strain, or heat transfer – are applied to predict how the object behaves when loads are applied. In sports engineering, Finite Element Analysis can be used to simulate how a golf club (Figure 8.2a) or tennis racket deforms when it hits a ball [107, 149, 208], how well a helmet absorbs energy during an impact, or how much pressure is exerted on different areas of a running shoe sole. Because Finite Element Analysis allows engineers to virtually "see inside" a product and analyse stress concentrations or deformation patterns, it is a useful tool for improving safety and performance.

Computational Fluid Dynamics focuses on the simulation of fluid flows – most commonly air or water – and is widely used when aerodynamics or hydrodynamics are important. In Computational Fluid Dynamics, the space around or within an object is divided into discrete grid elements, and the flow of fluid is calculated at each element using the Navier–Stokes and additional turbulence equations. Sports engineers use Computational Fluid Dynamics for various research purposes, such as analysing how swimwear influences water flow and the frictional and form drag, or how the shape and surface texture of a golf ball influences how air flows around it (Figure 8.2b), which can then be used to predict how it flies when struck. Computational Fluid Dynamics can

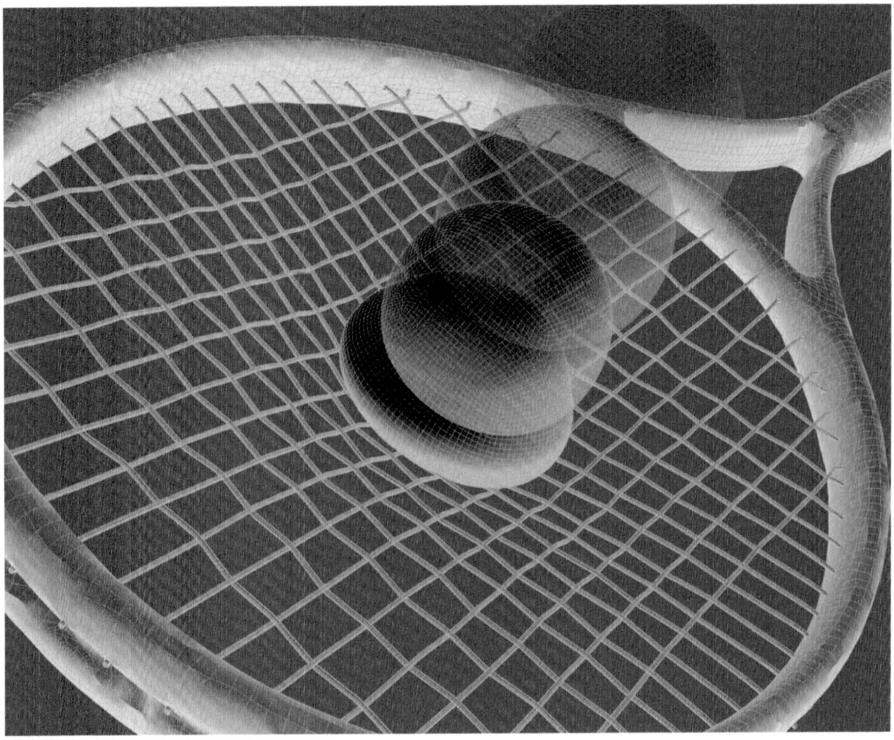

(a)

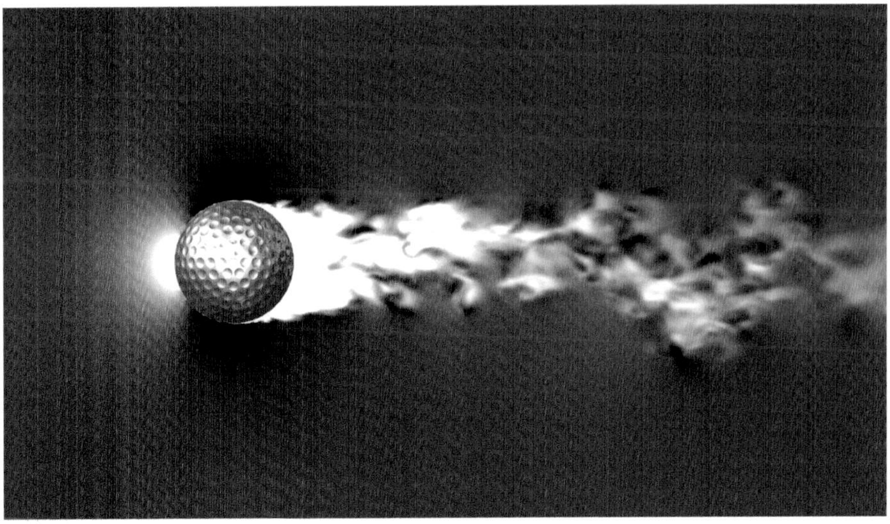

(b)

Figure 8.2 Images showing (a) a finite element model simulating an impact between a tennis ball and racket, and (b) a Computational Fluid Dynamics simulation of air flow around a golf ball.

also be used to inform product design choices, such as how best to reduce air resistance around a cyclist or ski jumper, or to optimise the airflow around a helmet or racing suit to reduce resistive forces or to improve cooling.

The use of virtual simulations in sporting goods design brings many benefits. One of the biggest advantages is efficiency. Physical prototyping is time-consuming and costly, especially if many design variations are tested. Virtual testing allows for rapid iterations and helps to identify the most promising designs early on. For example, optimisation tools within engineering simulation software allow sports engineers to efficiently find the best design for target inputs [209]. Virtual methods also offer deep insights. Simulations can provide detailed "full-field" information about stress, strain, pressure distribution, or airflow that would be hard or impossible to measure in real-world experiments. Another advantage is flexibility. Virtual models can easily be adapted to explore different sizes, materials, or usage scenarios, reducing the cost of physical prototyping.

Despite their benefits, virtual simulations have limitations. One of the most important challenges is that simulation results are only as good as the input data. If the material properties, loads, or boundary conditions used in a finite element model are inaccurate, the simulation will not reflect real-world behaviour. This principle is often summarised as "garbage in, garbage out". Another limitation is that models are often simplified to make them solvable. For example, complex joints may be fixed, dynamic movement may be ignored, or materials may be treated as perfectly elastic. Likewise, in fluids simulation it is still not economical or possible to accurately model turbulent flows in full. These simplifications can lead to results that miss important real-life effects. Virtual testing cannot replace real-world testing. While simulations are useful for exploring ideas and narrowing down design options, they must always be complemented by laboratory or field tests, or both, to validate the results. Real-life use involves complex interactions with athletes, variable environments, and long-term wear, all of which are difficult to capture in a virtual model.

Finally, computational cost can also be a barrier to engineering simulation. High-quality simulations – especially Computational Fluid Dynamics models – can require powerful computers and long processing times. However, emerging solutions such as cloud computing and distributed and graphics processing unit (GPU) processing are beginning to address this issue. These technologies allow engineers to run large-scale simulations more efficiently by outsourcing computation to scalable online platforms (e.g., [210]). As access to such infrastructure improves, virtual testing may become faster, more flexible, and more accessible – even for smaller companies and institutions. Many sporting goods brands benefit from virtual testing. In tennis, Finite Element Analysis has been used to simulate the dynamic behaviour of rackets during ball contact (e.g., [107, 149, 208]), helping to tune vibration characteristics and improve comfort. In cycling, Computational Fluid Dynamics is used to analyse airflow around the bike and rider [43, 44, 46], leading to more aerodynamic helmets, frames, and positions. In swimming and ski jumping, simulations help to optimise suit fit and surface texture to reduce drag and improve performance. These examples show how digital engineering tools are helping to shape the future of sport – from performance gear to protective equipment.

Box 8.3

Case example: Helmet development (simulation)

Using high-resolution 3D models, sports engineers simulate helmet impact performance with oblique angle impacts, showing how rotational forces affect brain injury metrics. Complementary Computational Fluid Dynamics simulations evaluate the aerodynamic flow over and through different vent configurations, estimating drag coefficients and temperature differentials at multiple wind angles. Simulated surface temperatures guide the refinement of internal air channels to optimise ventilation.

8.2.1 Mechanical laboratory testing

Mechanical testing in the laboratory is one of the most widely used and established methods for evaluating the performance of sporting goods [211]. In this controlled setting, sports engineers apply defined loads or movements to a product and measure its response. These tests help determine critical mechanical properties such as stiffness, shock attenuation, fatigue resistance, and rebound characteristics, depending on the intended application. An advantage of laboratory testing is its repeatability. Since the test conditions can be kept constant – including temperature, humidity, loading speed, and alignment – each sample can be tested under identical conditions. This allows for reliable comparisons between different products, prototypes, or material options. At the same time, the laboratory environment allows for high levels of control: disturbances caused by weather, user variability, or surface inconsistencies can be excluded, helping isolate the true material or product behaviour.

Mechanical testing is used throughout the sports product development process. In the early stages, material coupons or simplified geometries may be tested to evaluate material properties (see Chapter 5). Later, full-scale products – such as footballs, rackets, or protective pads – are subjected to more complex loading conditions. Nevertheless, the tests are used to analyse certain properties in isolation.

Various mechanical tests exist for sporting goods. Static tests assess how a product resists or deforms under slow, steady loads. These include compression, tension, torsion, and bending tests, which are often performed on universal testing machines. Dynamic or impact tests are used to simulate fast events like falls or collisions, such as drop tests used to assess helmet safety [212]. Fatigue testing involves applying repeated cyclic loading to simulate long-term use (see Chapter 5), which is especially relevant for footwear [213], bike components, skis, or training devices. Rebound tests assess energy return – a key factor in sports balls [102] and shoes [214].

Performing these tests often requires specialised machines and setups, collectively known as test rigs (Figure 8.3). These may include hydraulic or electromechanical actuators, drop towers, pendulums, or impact sleds. Measurement systems – such as load cells, accelerometers, or laser displacement sensors – are used to capture forces, deformations, accelerations, and time histories. High-speed cameras and digital image correlation systems [138, 215] can be used to analyse fast motions or strain

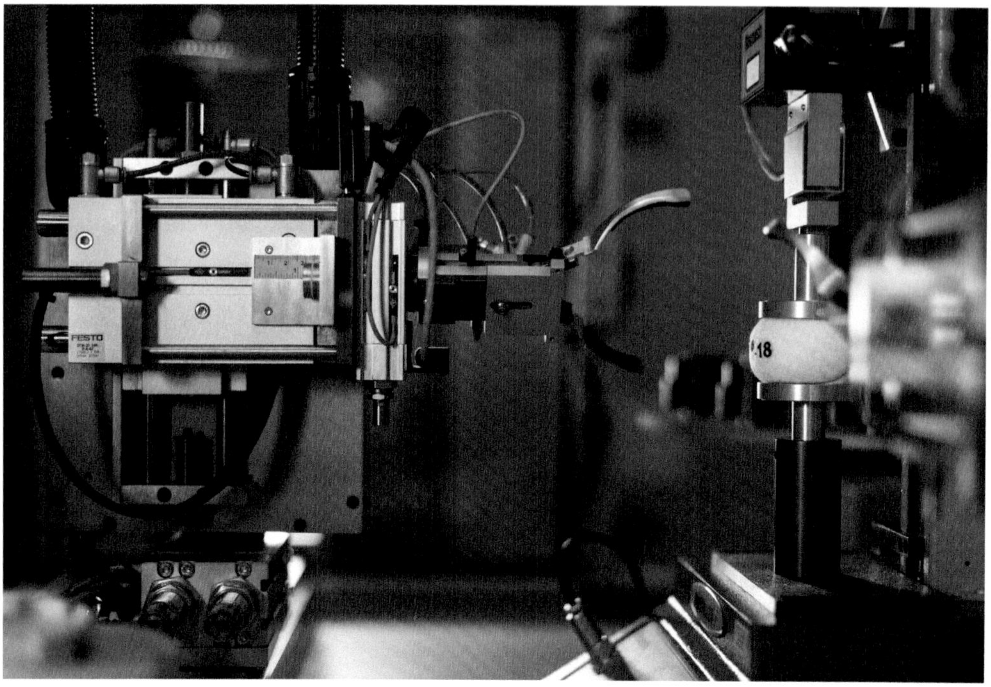

Figure 8.3 Obtaining mechanical tennis ball properties through compression testing. Copyright © This is Engineering.

fields on the surface of the object under test. These tools allow for detailed analysis of failure mechanisms, deformation behaviour, and performance under realistic loads.

A good test rig must be carefully designed to ensure that results are meaningful and reproducible. This includes proper alignment, secure and representative fixation of the test sample, and minimisation of unwanted degrees of freedom. For example, if a protective pad is not aligned correctly, the load may be distributed in an unrealistic way – resulting in misleading conclusions. Similarly, a poorly calibrated force sensor or high-friction guides may distort results. The role of the sports engineer is not only to operate machinery but to plan and execute meaningful tests, select appropriate test methods, analyse results, and understand their limitations. They must be able to interpret unexpected results, question anomalies, and, when needed, redesign test setups to better reflect real use cases.

Although laboratory tests offer high accuracy and control, they are always abstractions of reality. No athlete moves exactly like a test machine. Equally, when designing and selecting test machines and associated protocols, sports engineers must balance realism with factors such as complexity, cost, space, and repeatability. Therefore, the data from laboratory testing must be interpreted in context – often in combination with virtual simulations or field data. Mechanical testing remains the foundation of product development and regulation in sport, offering a bridge between material properties and athlete experience. With a growing range of available tools, from traditional load frames to advanced optical systems, the challenge today is not just to measure – but to measure meaningfully.

Box 8.4

Example procedure for the design of a mechanical test (systematic development or adaption of suitable test methods)

Like many sports engineering development activities, this is an iterative process. This means it is possible to return from higher levels, and individual sub-steps are often completed several times.

1. Define the purpose of the test (methods: benefit–risk assessment ISO 4980 [205], market analysis) within a test strategy, e.g., footwear: shock attenuation, traction, torsional stability, flexural stability, etc.
2. Clarification of the state of the art (determine whether tests developed for other purposes can be used or modified)
3. Analysis of direct or indirect interactions between the sporting goods and the human body, as relevant for the test
4. Closing knowledge gaps identified in step 3 by means of measurements or literature
5. Specification of test statistics, measurands, and limit values
6. Design and build the test rig
7. Determine whether the requirements of the development are met, with modifications if necessary
8. Determine test conditions and test protocols
9. Conduct pilot study
10. Characterisation and validation of the test method (method: Measurement System Analysis for accuracy, repeatability, reproducibility)
11. Preparation of work instructions for the operating personnel
12. Release of the test bench and procedure

Box 8.5

Case example: Helmet development (mechanical laboratory testing)

Sports engineers conduct drop tests at multiple impact locations and velocities using an anthropomorphic headform instrumented with tri-axial accelerometers. Results show the peak acceleration remains below critical thresholds at all test points. Additional mechanical tests assess strap retention and buckle strength, as well as airflow resistance through the helmet's vents. Air permeability measurements help estimate potential for convective cooling and inform subsequent simulation steps.

8.2.2 Human-based laboratory testing

While mechanical and virtual testing methods are essential for evaluating the physical behaviour of sporting goods, they do not provide a complete picture. The most important questions are rarely related to how a sports product performs in isolation, but rather how it behaves when used by an athlete. This is the focus of biomechanical testing.

Figure 8.4 Photo showing an example of biomechanical experiments during running. The positions of markers placed on the legs can be tracked with an optical motion capture system.

Biomechanical testing investigates how sporting goods interact with the human body. It aims to understand the loads, motions, and pressures involved during real movements, and how these are influenced by design choices. Such testing plays a vital role in improving both performance and safety and helps sports engineers create products that are not only technically advanced but also "athlete-friendly".

There are two main approaches: human-based testing, which uses real athletes (see refs. [216–218]), and surrogate-based testing, which uses physical models to simulate parts of the human body in controlled and repeatable ways. In human-based testing, athletes perform sport-specific movements in a laboratory while using the equipment under investigation. Sensors and cameras capture their motion, force distribution, muscle activity, or pressure profiles. Commonly used systems include motion capture with reflective markers and high-speed cameras, which allow tracking of joint angles and segment positions in three dimensions (e.g., Figure 8.4). Force platforms or load cells can measure ground reaction forces during running, jumping, or landing. Pressure measurement systems placed in insoles or on the surface of contact zones can reveal how load is distributed under the foot [219, 220], inside a ski boot [18], or across a saddle.

Increasingly, equipment itself is being instrumented, e.g., tennis rackets equipped with accelerometers, ski bindings with strain gauges, or mountain bike handlebars fitted with vibration sensors (e.g., [18]). This allows for the direct measurement of how equipment deforms or vibrates during real use. In addition, wearable sensors such as inertial measurement units (IMUs) make it possible to monitor body movement and loading in training or competition environments, far beyond the laboratory.

Human-based testing has many strengths. It offers high ecological validity and allows direct insight into user response, comfort, and technique. Ecological validity is essentially how well the study relates to the real-world setting, which for sporting goods is when used for sport. Human-based testing also brings challenges. Human movement is inherently variable – even the same athlete will not move exactly the same way twice. Different body sizes, techniques, and levels of fatigue introduce complexity, making direct comparisons difficult. The time and effort associated with planning and running a study with suitable participants [221], which includes recruiting them, should also not be overlooked. In addition, some tests may be unfeasible or unethical (see ref. [222]) to perform with human participants, particularly in high-risk scenarios.

To overcome these limitations, surrogate-based testing is used. In this approach, parts of the human body are represented by physical models that mimic their size, shape, and mechanical properties. These may range from simple metal cylinders to complex anthropomorphic test devices (ATDs), also known as "crash test dummies". Surrogates are typically used when repeatability and safety are critical [10, 145, 223–229] – such as in impact testing.

For example, helmet testing often involves dropping a head form equipped with internal accelerometers onto a hard surface to simulate a fall. The measured acceleration is used to assess how well the helmet protects the head. In wrist protector testing, the device is mounted on a forearm-shaped surrogate and struck by a falling weight [224, 230]. The peak force transmitted through the protector is measured using a load cell. These surrogate tests form the basis for many standardised protocols, as they allow precise control of conditions and comparison between different designs.

Biomechanical testing – both with humans and surrogates – adds a valuable dimension to sports engineering. It allows researchers and designers to evaluate not only how strong or light a product is, but how it affects movement, comfort, and injury risk. It supports the development of more effective, user-friendly equipment, and helps tailor designs to the needs of different user groups. As wearable technologies and mobile sensing continue to advance, biomechanical testing is becoming increasingly mobile and accessible (see ref. [18]). Data can now be collected outside the laboratory – in the field, during training, or even in competition – and used to inform design decisions in real time. This development is helping to close the gap between research and practice and is opening new opportunities for athlete-centred design in sport.

Box 8.6

Case example: Helmet development (biomechanics)

Ten trained cyclists complete 30-minute high-intensity sessions with different helmet prototypes on a cycling ergometer under fan-simulated wind. The test order of the prototypes is randomised between participants, who rest between trials, so fatigue does not influence the results. Electromyography (EMG) data reveals reduced neck muscle activity with the new design, likely due to weight redistribution. Infrared thermography and in-helmet temperature sensors document local heat accumulation. Subjective ratings by the participants indicate reduced sweat accumulation and improved airflow, particularly in the forehead region.

8.2.3 Field testing in sport environments

Field testing plays a vital role in evaluating sporting goods under real-world conditions. Unlike laboratory or simulation-based methods, which focus on controlled and often simplified scenarios, field testing is conducted where the sport takes place – on the court, the ski slope, the road, or the pitch. It provides insight into how sporting goods perform during real movements, in natural environments, and under authentic physical and psychological demands. This is why sporting goods brands seek feedback on late-stage prototypes from high-level athletes, as discussed in Chapter 2.

The strength of field testing lies in its ecological validity. While laboratory tests offer high repeatability, they cannot fully replicate the unpredictable and dynamic nature of actual sport. Field testing captures how equipment behaves when used by athletes in real motion patterns, over varied surfaces, in changing weather conditions, and often under fatigue or competition pressure. To collect useful data under these conditions, modern field testing relies on compact, mobile, and often wearable technologies. IMUs, for example, can be attached to the body or to equipment to track acceleration, angular velocity, and orientation in three dimensions [231]. These sensors are commonly used to analyse running gait, cycling dynamics, or ski turns. Satellite-based navigation systems help track position, speed, and movement patterns during outdoor sports such as football, cycling, or cross-country skiing. Pressure insoles placed inside footwear or ski boots measure how forces are distributed across the foot during ground contact or during turns [18]. Force-sensing pedals in cycling allow for the measurement of left–right balance and pedalling efficiency.

Other tools include instrumented mouthguards [232] or helmets with embedded IMUs, which are used to detect head impacts in contact sports. Mobile apps and video-based analysis software are also widely used in field testing, especially in combination with subjective ratings and athlete feedback. In addition to objective measurements, perception testing is often used during field evaluations to capture how sporting goods "feel" and perform from the athlete's perspective. It provides subjective data on comfort, control, usability, or confidence, which may not be reflected in physical measurements alone. Perception testing methods include structured interviews, Likert-scale questionnaires [233], or comparative rating tasks and are typically done during or just after use. Such user-centred feedback is especially valuable when designing equipment for comfort, ease of use, or enjoyment – for instance in footwear, rackets, or protective gear. It can also reveal relationships between measurable performance and perceived effectiveness (e.g., [190, 191]). Including subjective assessments alongside sensor-based data helps ensure that field testing reflects not only physical function but also real-world user acceptance and experience.

Box 8.7

Perception testing in field environments

Perception testing methods commonly used in sporting goods evaluation include:

- **Likert scales** (e.g., 1 = very uncomfortable to 5 = very comfortable)
- **Paired comparisons** (e.g., "Which of these two shoes felt more stable?")
- **Open-ended questions** (e.g., "Describe how the handlebar grip felt during rough sections")
- **Semantic differentials** (e.g., "The racket felt: ☐ light – ☐ heavy")

Typical questions focus on:

- Perceived **comfort** during use (e.g., pressure points, heat build-up)
- Perceived **control or stability** (e.g., on uneven surfaces or during turns)
- Confidence or **safety perception** (e.g., under impact or at high speed)
- General **preference** or willingness to use the equipment again (e.g., "Which of these two snowboards do you prefer?")

Such responses are often collected just after use in the field, to ensure relevance and authenticity.

Field testing is widely applied across many sports. In running, pressure insoles combined with IMUs can provide estimates of ground reaction forces and identify how a shoe design affects foot strike patterns or pronation. In skiing and snowboarding, pressure and vibration sensors embedded in boots or bindings help sports engineers analyse turning technique, edge control, and shock exposure. In cycling, accelerometers on handlebars and saddles measure vibration exposure on different surfaces, while sensors on the suspension track the response to impacts and terrain changes. In team sports like football or rugby, outputs from wearable sensors can be used to calculate metrics like total "load", sprint effort, and even collision intensity.

An advantage of field testing is that it enables direct observation of how sporting goods are used in practice. It supports athlete-centred product development and allows coaches and scientists to validate the assumptions made in laboratory studies. It can also reveal behaviours or effects that are difficult to predict or detect in controlled settings – such as slipping, compensatory movements, or the influence of psychological factors.

Field testing also presents challenges. Environmental variability makes it hard to repeat tests under identical conditions. Data collection and quality may be affected by movement artefacts, sensor placement, or limited battery life [234]. The complexity of real-world motion and the volume of data collected can complicate analysis and interpretation. Despite these challenges, field testing remains a crucial complement to laboratory testing, helping to close the gap between experimental results and real-world performance.

With advances in sensor technology, data processing, wireless communication, and machine learning, field testing is becoming more accessible and precise. Commercial sporting goods are increasingly including integrated measurement tools – from navigation systems in watches to smart insoles and artificial intelligence-based feedback systems. These technologies are not only used by elite athletes, but increasingly by recreational users and sports engineers developing products. Such data collected routinely by individuals and teams during sport presents new opportunities for sports engineers, if it can be shared and accessed ethically.

Looking forward, data collected during field use may increasingly inform virtual models, improve standardised laboratory protocols, and enable adaptive product development. This shift towards integrated, user-informed testing represents a step towards more personalised and evidence-based sports engineering.

Box 8.8

Case study: ISO 8783:2015 – Guidelines for conducting slope performance tests for alpine skis

ISO 8783:2015 [235] outlines a systematic approach to assess the user perception of alpine skis. Key performance characteristics of a ski subjectively evaluated by skiers in the tests include the following parameters:

Turnability refers to how easily it initiates and completes a turn, with minimal force or effort.

Edge grip assesses its ability to maintain a firm hold on hard, icy surfaces.

Steering behaviour evaluates its responsiveness and precision when changing direction, including how smoothly it transitions in and out of turns.

Stability describes how well it maintains its intended direction under external forces, such as on uneven terrain or at high speeds.

Quietness, or damping behaviour, measures how effectively it absorbs shocks and vibrations, ensuring a smoother ride.

Skier friendliness captures how intuitive and manageable it feels.

To ensure valid comparisons between skis, tests are conducted on groomed slopes under consistent conditions. The use of reference skis provides a benchmark for testers to rate others against. Results are recorded on a five-point Likert scale [233] ranging from "Very Good" to "Very Poor". Testers must be advanced skiers and remain physically consistent throughout the testing process. A committee of technical experts oversees the tests, ensuring adherence to the standard.

Benghanem et al. [236] used a similar approach when categorising alpine skis. Studies have shown that engineering properties of skis, such as mass and shape, torsional and bending stiffness, and camber length and height, influence perceived slope performance [237–240]. Following ISO 8783's standardised field testing protocol allows sports engineers to get a better understanding of these interrelationships. However, running such a test protocol requires considerable planning, effort, and resources, which may be unfeasible for many ski brands.

Box 8.9

Case example: Helmet development (field testing)

To evaluate ventilation performance under real-world conditions, cyclists wear the prototype helmet during extended outdoor training rides. Small wireless temperature sensors are placed inside the helmet (e.g., at the forehead, crown, and rear) to monitor heat build-up in different airflow zones. Data are recorded continuously while athletes ride on varied terrain and at different intensities.

Ambient temperature and wind conditions are logged in parallel using on-bike weather sensors. Riders complete standardised post-session questionnaires to assess subjective thermal comfort, sweat accumulation, and airflow perception.

Results indicate that the new helmet design reduces peak internal temperatures by a mean of 1.6°C over the previous model, particularly during high-speed segments. Subjective ratings further indicate improved cooling and overall ventilation effectiveness.

8.2.4 *Testing standards and governing body regulations*

In many sports, the use of sporting goods is governed by a combination of regulations and standards. Regulations are binding rules issued by international sports federations or governing bodies. They define which types of sporting goods are permitted in training and competition and often set minimum safety and performance requirements. Standards, in contrast, are detailed technical documents that describe how equipment should be tested, measured, or constructed. They are developed by expert committees within organisations such as the International Organization for Standardization (ISO) or American Society for Testing and Materials (ASTM) International.

While compliance with a standard is often voluntary, it may become mandatory if referenced in a regulation or adopted by a sport's governing body. Before a product – such as a football, ski binding, or helmet – can be approved for competitive use, it must often demonstrate conformity with both relevant regulations and applicable standards. The standards define not only what must be tested, but also how the tests must be performed – including details such as test setup, equipment configuration, loading conditions, and pass criteria. By following these specifications, test procedures become reproducible, transparent, and comparable across different laboratories and sporting goods brands worldwide.

A well-written standard provides a consistent and objective framework for testing. It defines the scope and purpose of the test, the required mechanical setup, the measurement methods, and pass thresholds. For sports engineers and sporting goods brands, this facilitates product development and certification in a structured and verifiable manner. For athletes and sport authorities, it helps ensure that sporting goods meet basic expectations for safety and fairness – especially in elite competitions where small differences in equipment can influence performance outcomes. A key benefit of standardised testing is safety assurance. For example, helmets, protective pads, or wrist protectors must demonstrate sufficient force limitation in impact tests. Products that do not meet these criteria are not allowed in competition. Standards also ensure that products from different brands can be directly compared – an essential aspect in regulated sports or in procurement processes.

Standardised testing is not without criticism. Many standards are based on simplified or outdated assumptions. For example, impact tests are often performed with fixed metal anvils, which do not reflect the biomechanical behaviour of the human body [241]. Similarly, some tests assume a single linear loading condition, even though real sports impacts are multi-directional. Standards may also focus on minimum requirements, rather than encouraging best-in-class performance. This means that products that just meet a standard will not necessarily perform well during use and may simply meet the minimum requirement for the task. Some sporting goods brands will design their products to exceed the minimum requirements of corresponding standards. There are also some independent rating systems for sporting goods, such as the *Virginia Tech Helmet Ratings* and the *Helmet Impact Protection Effectiveness Rating* (Hiper), which through transparent reporting of results may encourage brands to improve their products.

Moreover, the process of developing or updating standards can be slow. International coordination, consensus processes, and political considerations often delay the adoption of improved methods. This can make it difficult for innovative products to gain approval, especially when their benefits fall outside the scope of current standards. For sports engineers, this presents a challenge: they must follow current protocols while recognising their limitations and advocating for improvements where necessary.

Box 8.10

Case study: A standard for snowboarding wrist protectors

An example of a standardised test procedure is ISO 20320:2020 [202] for wrist protectors in snowboarding. This test simulates a fall on the palm of a hand by placing the wrist protector on a fixed metal hemisphere and dropping a weight onto it. A load cell records the peak force transmitted through the protector. The wrist protector must reduce the peak force below a specified threshold to pass. The procedure is simple, reproducible, and well-defined. However, questions remain about its realism: does a rigid hemisphere truly mimic the human wrist? Does the fall scenario match the typical movements in snowboarding? These concerns highlight the need to regularly revise standards based on new scientific evidence and practical experience.

Ultimately, standards and regulations serve an important function – but they must develop with the sport, the equipment, and the available technologies. Combining laboratory tests with real-world data and virtual simulation offers new opportunities to develop more accurate, relevant, and athlete-centred testing protocols. The role of sports engineers in this process is to go beyond compliance and contribute to the next generation of meaningful, realistic, and inclusive testing standards.

Box 8.11

Case example: Helmet development (standards)

The development team consults EN 1078:2012 [206] and ASTM F1447 [242] protocols to design the test matrix. Headforms, drop heights, and impact surfaces are chosen accordingly. A pre-certification test confirms the helmet meets all required thresholds.

8.3 Experimental considerations in sports engineering

Conducting meaningful and trustworthy experiments is at the heart of sports engineering. Whether in a laboratory, a virtual simulation, or a field test, engineers must ensure that the data they collect is reliable, repeatable, and relevant. To do so, they must consider various experimental factors – from test planning and measurement accuracy to statistical analysis and results interpretation.

A key requirement is that an experiment must provide results that are both correct and consistent. This means not only that the measurements reflect the actual physical behaviour of the product (validity), but also that they would be the same if the test were repeated (repeatability and reproducibility). Furthermore, the outcome should have practical significance: would it make a difference to athletes, coaches, or governing bodies?

8.3.1 *Uncertainty, accuracy, and precision*

All measurements are subject to uncertainty. This may arise from inherent variation between products, differences in how repeated tests are performed, or limitations of the measurement tools. Reducing uncertainty increases confidence in test results. For example, when measuring the mass of tennis rackets, even those of the same model may vary slightly due to manufacturing tolerances. The more precise the weighing scale and the more consistent the manufacturing, the smaller the measured variation will be.

Accuracy refers to how close a measurement is to the true or accepted value [243]. Precision describes how close repeated measurements are to one another [243]. Ideally, a test should be both accurate and precise – with measurements that cluster tightly around the correct value. In practice, accuracy is influenced by measurement device calibration and setup, while precision is typically affected by random error.

To reduce, or account for, error, it is essential to control, or at least record, environmental conditions such as temperature and humidity. Measurement tools should be selected carefully, maintained, and calibrated regularly [18, 220, 232, 244, 245]. Commercial systems often come with calibration certificates, but sports engineers may also need to calibrate custom devices themselves. Using reference systems, such as motion capture systems or certified force plates, helps verify the performance of newer or mobile technologies such as wearable sensors.

Box 8.12

Case example: Helmet development (uncertainty)

Accelerometers are calibrated each year. The standard deviation of repeated measurements is used to estimate uncertainty. Results are reported with 95% confidence intervals.

8.3.2 *Repeatability and reproducibility*

Repeatability means that a test produces similar results when the same person repeats the experiment under the same conditions. Reproducibility, by contrast, refers to whether other people can reproduce the results using the same method, either in the same or another laboratory. Achieving reproducibility requires transparent documentation, clear test protocols, and attention to experimental design. It is critical in standardised testing, where results must not depend on the test location or personnel. A common approach to assess reproducibility is the round-robin test: different laboratories perform the same test independently on the same sample type, and the results are compared. This process helps identify inconsistencies and can inform decisions on refining the test procedure to improve reproducibility.

Box 8.13

Case example: Helmet development (repeatability)

Each prototype is tested three times at two different facilities. Results show high repeatability (coefficient of variation <3%) and good inter-lab agreement.

8.3.3 Statistical analysis and reporting

Once measurements are collected, appropriate statistical tools must be applied to interpret the results. Descriptive statistics such as mean, median, mode, standard deviation, and range help summarise the data. More advanced tools, such as z-scores or coefficients of variation, allow for comparison across different contexts or units.

Outliers – values that differ strongly from the rest – can be detected using z-scores and visualised through boxplots or histograms; whether to include or exclude outliers should be justified, not arbitrary. The repeatability coefficient, derived from standard deviation, can help quantify how consistent repeated measures are. In hypothesis testing, significance levels (typically $p < 0.05$, i.e., <5%) are used to assess whether an observed difference is likely due to chance.

Measurements should always be reported with a level of detail that reflects their accuracy. For example, if a ruler with millimetre divisions is used to measure the maximum width of a ski, reporting results in centimetres or micrometres may be misleading. Careful reporting also includes specifying the number of repetitions, the sampling strategy, and any assumptions made during data analysis.

Box 8.14

Case example: Helmet development (statistics)

Temperature data from ten athletes is analysed using repeated-measures ANOVA to evaluate the effect of helmet design on internal temperature. A significant main effect ($p < 0.05$) is observed. Post-hoc comparisons indicate that the new helmet results in a mean of 1.4°C lower peak temperature and significantly higher comfort ratings on a 7-point Likert scale. Correlation analysis reveals a strong association between measured forehead temperature and subjective cooling perception ($r = -0.78$).

Box 8.15

Case summary: Helmet development

Through mechanical, virtual, and field-based testing, the helmet achieves EN 1078:12 certification, lowers aerodynamic drag by 12%, and improves thermal comfort in training. The project exemplifies best practices in test planning, execution, and interpretation in sports engineering.

8.4 Chapter summary

This chapter provided a broad introduction to testing sporting goods. It outlined the vital role testing plays in sports engineering by ensuring that sporting goods are safe, functional, and fit for purpose. The various complementary methods available – including mechanical laboratory testing, computer-based simulation, biomechanical analysis, and field testing under real-world conditions – were summarised. This included a discussion on the strengths and limitations of each method. It was also noted how, in practice, these methods are often combined to obtain a comprehensive picture of sporting goods performance. It was acknowledged that mechanical tests offer precision and repeatability, while virtual methods enable rapid iteration and early design optimisation. Biomechanical and perception-based testing help evaluate athlete–equipment interaction, and field testing ensures ecological validity in dynamic environments. Standardised protocols set by governing bodies ensure that testing is comparable and transparent, but they must develop with technology and user needs. Emerging technologies such as wearable sensors, cloud computing, and artificial intelligence-driven analysis are expanding the scope and flexibility of testing in sport. These developments open the door to more personalised, adaptive, and efficient test strategies – both in elite performance and product development. For future sports engineers, understanding and applying the right test methods is essential for improving safety, performance, and user experience in sport. There is more to learn about the testing of sporting goods than the basics covered here, some of which can be found in the suggested further reading.

8.5 Suggested further reading

The following suggestions for further reading go deeper into the topics introduced in this chapter:

- *Routledge Handbook of Sports Technology and Engineering*, 1st and 2nd editions [10, 86]
- *Sports Biomechanics: Reducing Injury Risk and Improving Sports Performance*, 2nd edition [218]
- *Wearable Sensors in Sport: A Practical Guide to Usage and Implementation,* 1st edition [231]
- "ISO 5725-1:2023 – Accuracy (trueness and precision) of measurement methods and results – Part 1: General principles and definitions" [243]
- *21st Century Sports* [246]
- *The Science and Engineering of Sport Surfaces* [247]
- "FIFA Quality Programme for Footballs (outdoor, futsal and beach soccer footballs) Test Manual" [102]

9 Gauging the effect of technology on sport

This chapter starts by outlining the importance of gauging the effect of technology on sport, while discussing the challenges of doing so. Different approaches are presented and discussed, including modelling, and the use of large datasets populated by data now routinely collected by equipment brands, sports teams, and at sporting events. Examples for modelling include those used to simulate the trajectories of sporting projectiles and to predict the time for a cyclist to travel a set distance. Such models can be used to predict the effect of equipment design on performance, with implications for equipment development and regulation, and athlete training and strategy. The limitations of predictive models are also noted.

Data is now routinely collected at sporting competitions and events via optical motion capture systems and sensors worn by participants. The advantages of using such data as a tool to gauge the effect of changes in equipment are presented. Examples for big data include analysing parameters that influence the flight of projectiles [248] in sports such as tennis, cricket, and baseball, and the effect of running shoe choice on marathon times. Further discussion on the ethical considerations of using engineering and technology to enhance sporting performance, and influence officiating and fan experience, is included.

9.1 The importance of gauging the effect of technology on sport

There are various reasons why it is important to gauge and understand the effect of technology on sport. From a scientific perspective, it is interesting to learn and understand how technological developments can influence sport and enhance human performance. Effective communication of such knowledge can help raise public interest in sport and technology and may even encourage people to pursue careers in science, technology, engineering, art, and maths (STEAM). From a practical perspective, application of knowledge can benefit sport and the sports engineering industry. Sports engineers can also use such knowledge to better support disabled athletes, while ensuring fair competition.

Regarding the sporting goods industry, a better understanding of the types of developments that enhance performance can help sports engineers focus effort and resource to where it can be most effective, bringing the biggest benefits and return on investment. Such knowledge can also feed into evidence-based marketing and sales strategies and potentially help match consumers and end users to appropriate products. This can in turn enhance enjoyment and safety and increase participation in sport, all with associated health benefits. Knowledge of how to use technological developments to enhance safety, such as the appropriate use of new materials to develop better helmets, is particularly important. At the elite level, developments in sporting goods and facilities can enhance athlete performance and hence the spectacle of sport.

DOI: 10.4324/9781003411086-9

From regulatory and officiating perspectives, better knowledge of the effect of technology on sport can help sporting governing bodies to set evidence-based rules and regulations for sports equipment. This can help governing bodies to be proactive rather than reactive to technological developments. For example, the Royal and Ancient (R&A) Golf Club,[1] United States Golf Association (USGA),[2] and the International Tennis Federation (ITF)[3] have advanced technical centres and test facilities, which they use to monitor developments in associated sports products, and make informed regulatory decisions [36, 249, 250]). Such a proactive and evidence-based approach can help contribute to fair competition and avoid the need for urgent rule changes to account for "surprising" technological developments. Keeping abreast of all technological developments, predicting their potential effect on sporting performance, and making appropriate and timely regulatory decisions is not trivial for a governing body, however. For example, the turn of the 21st century saw the emergence of full-body polyurethane swimsuits (e.g., Figure 9.1a) [42, 110]. These swimsuits were coined "super-suits", and they helped athletes break 25 records at the 2008 Beijing Olympics [111]. This prompted the Fédération Internationale de Natation (FINA) (now World Aquatics) to tighten their rules around swimsuits in 2010, in what is known as the "Dubai Suit Charter". More recently, developments in distance running shoes (e.g., Figure 9.1b) led to improvements in marathon running times, and subsequent rule changes by World Athletics [56].

9.2 Challenges in assessing the effect of technology on sport

Gauging the past, current, and future effect of technology on sport is not trivial. There are challenges associated with evaluating the influence of a technology that stem from a broad array of in-field variables, and the available data collection and analysis techniques. A major factor is the definition of sports technology, or how best to define sports technology. For example, is "sports technology" restricted to sports equipment and clothing used by athletes, or does it include things such as sports facilities, training and feedback tools, and injury recovery devices? Specifically, where the line is drawn will ultimately influence the estimated effect of technology on sport. Linked to this is how "sport" is defined. Is this all sport together, or specific sports?

Some sports have been more influenced by technology than others. This is partly because sporting governing bodies differ in how they allow and embrace technology. Cycling, for example, relies predominantly on technology in the form of a bike, and the sport has changed and expanded considerably over the years. Technological developments have led to new cycling disciplines such as BMX and mountain biking (see Figure 5.10 and Figure 6.3). Indeed, technology can lead to new sports, such as snowboarding, and for this reason the effect of technology on sport participation and uptake should not be overlooked. For some sports, such as tennis [29, 30, 33, 34, 36], the incremental, and occasional step change, improvements in equipment are clear. In contrast, skateboarding was revolutionised by the introduction of polyurethane wheels in the 1970s, with relatively little change since beyond some variation in the shape of the classic board design. Regardless, debates and analysis on the effect of technology on sport tend to focus on the performance of elite athletes, particularly Olympic and world records. The remainder of this chapter will hence focus on elite sports performance.

Isolating the impact of technology on elite sporting performance can be difficult. Athletes differ and respond differently to technological developments; as such, whereas some improvements in equipment may have obvious benefit, others may have a varying

(a)

Figure 9.1 Photos showing examples of advanced sporting goods that have enhanced athlete performance, (a) swimsuit and (b) modern distance running shoes.

(b)

Figure 9.1 (Continued)

influence on the level of observed performance enhancement. Other factors besides technology can also contribute to improvements in sporting performance and new records. These can include better training, talent identification, diet, recovery and medical support, and even performance-enhancing drugs. There is also increased competition driven by financial incentives and the number of nations competing. For example, such inherent improvement and progression in sporting records can be seen in quicker running times for different distances, despite the sport's relatively low reliance on technology. It is the complex and conflicting combination of all these influencing factors that make it hard to isolate specific variables and hence quantify the exact effect of technology on sporting performance.

9.3 Using historical records to gauge the effect of technology on sporting performance

Sports engineers often look at the progression of world records when attempting to gauge the effect of technology on sporting performance (e.g., [43–46, 110, 251–254]). Figure 9.2 shows how the Union Cycliste Internationale (UCI) cycling hour record distance has progressed since the 1890s to present. The UCI hour record is the longest distance cycled in an hour from a stationary start. As the race time is one hour, the average speed in km/hr can be obtained by simply dividing the distance in km by one, i.e., the same values as distance, just different units. The hour record is a nice example to showcase how equipment developments can improve performance whilst also demonstrating the challenges of quantifying improvement. The record also clearly demonstrates the considerations and challenges that can be posed by regulatory bodies. Figure 9.2 clearly shows that the record distance has increased over time. This progress is in part due to improvements in bike design (see Figure 9.3), as well as other factors already noted, such as diet, training and rider physique.

The introduction of the UCI hour record around the turn of the 20th century was a period of rapid progress, with growing interest in the burgeoning sport coinciding with early innovation in bike design. A lack of progress during both world wars and their aftermath is also clear. The record began to progress again from around the mid-1950s until 1972. In 1972, Eddy Merckx came close to achieving 50 km in the high-altitude Mexico City, with its lower air density and hence lower aerodynamic drag (see Chapter 4). This new record remained for over ten years until it was broken in 1984 with a bike featuring disc wheels to reduce aerodynamic drag. This marked the start of a new period with the 1990s seeing rapid progress in the record due to innovations in equipment and riding positions, including the infamous "superman position".

Due to concerns over how technological developments in the 1990s were causing such rapid progress, the UCI changed their rules in 1997. All attempts since 1972 were reclassified as "Best human effort", and from then on, any rider attempting the UCI hour record must use a bike of traditional design (e.g., Figure 9.3a). Following this the record was only broken twice, in 2000 and 2005, until another rule change in 2014. During this 1997–2014 period the maximum distance remained below that of the lowest "Best human effort", indicating stagnation. The 2014 UCI rule change introduced the use of

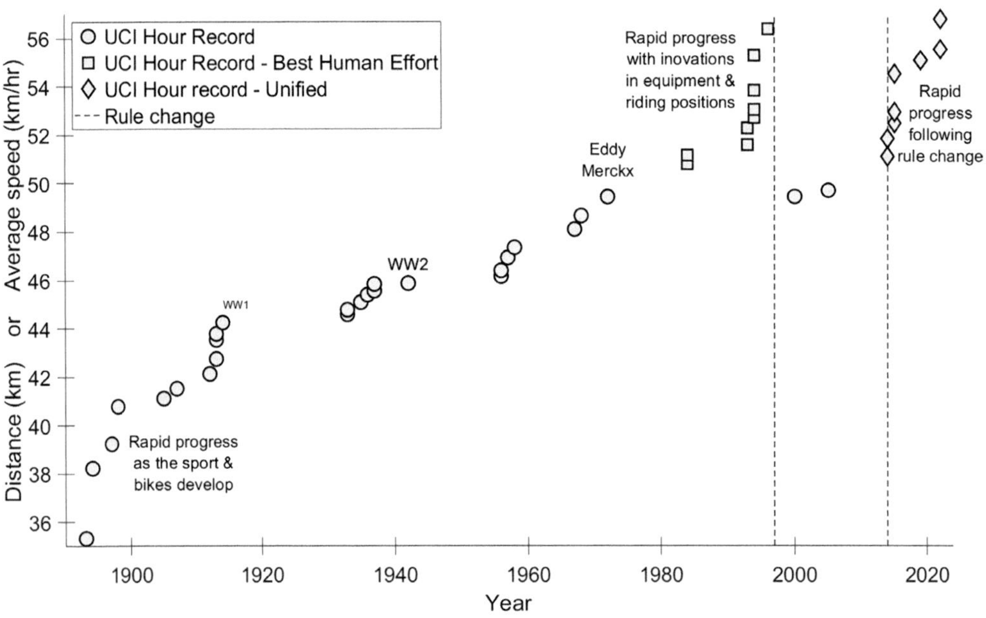

Figure 9.2 Graph showing how the cycling hour record distance (and average speed) has progressed since the 1890s to present. Periods of slow and fast progression of the record can be seen, corresponding to both world wars, and progression of cycling equipment and rule changes. Data from Wikipedia and cross-checked against www.wolfgang-menn.de/hourrec.htm.

(a)

(b)

Figure 9.3 Photos showing racing bikes, (a) traditional and (b) modern.

endurance and track bikes. The "Unified UCI Hour Record" included the original hour record distances (i.e., it continued from 2005) but not those of the "Best human effort". This rule change caused another period of rapid progress, with the current record set in 2022 just exceeding that of the "Best human effort".

Looking at such progress in the UCI hour record clearly shows that technological developments can enhance sporting performance. It also highlights the importance of managing technological developments appropriately. However, due to other factors, such as global events, the general interest in the sport at different time periods, and better sports science support, accurately quantifying the effect of technology on sporting performance in this way is not possible. There is only a limited amount of data available, with gaps around some periods, and the time athletes had to adapt to new technology is hard to gauge.

The next subsections of the chapter will focus on other techniques for estimating the effect of technology on sporting performance, namely modelling and big data. Using a combination of such techniques, the sports engineer can better gauge the effect of technology on sport.

9.4 Modelling approaches for assessing the effect of technology on sport

Sports engineers can develop and use models to estimate the effect of technology on sporting performance. This subsection introduces the concept of using models for estimating sporting performance. Examples from different sports and sporting scenarios are described, including cycling, running, tennis, and ball flight. The balance between model complexity and accuracy is discussed. In this context, Haake [45] introduced the concept of a performance improvement index. Much of this subsection builds on the foundations introduced in Chapters 3 and 4.

9.4.1 *Basic cycling model*

We start by returning to the example of the hour cycling record. Figure 9.4 shows the main forces acting on a cyclist. These forces can be used as the basis of a model for predicting cycling performance (e.g., [45, 251, 255, 256]). Equation 9.1 is an expression of Newton's second law of motion, as introduced in Chapter 3, for a cyclist (bike and rider):

$$F = m_c a_c = F_a - F_r \qquad \text{Equation 9.1}$$

where F is force, m_a is the cyclist mass, a_c is their acceleration, and the subscripts a and r refer to applied and resistive, respectively. As shown in Figure 9.4, the resistive forces are air resistance (i.e., aerodynamic drag, F_d) and friction (i.e., rolling resistance, F_{rr}). Substituting these resistive forces into Equation 9.1 gives:

$$m_c a_c = F_a - F_d - F_{rr} \qquad \text{Equation 9.2}$$

We can then see that the acceleration of the cyclist can be obtained by dividing the right-hand side of Equation 9.2 by their mass:

$$a_c = \frac{\left(F_a - F_d - F_{rr}\right)}{m_c} \qquad \text{Equation 9.3}$$

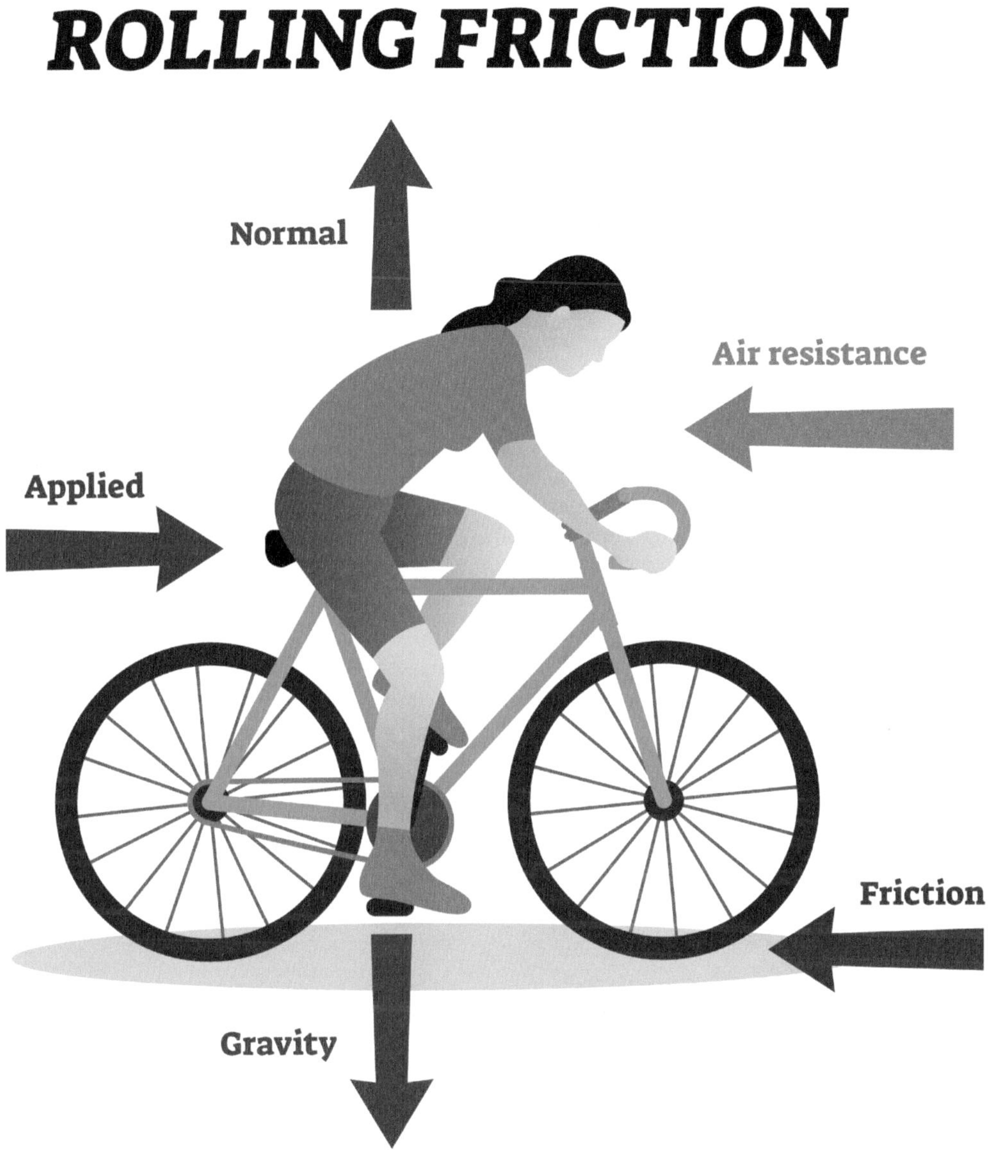

Figure 9.4 Illustration showing the main forces acting on a cyclist riding on a flat straight. The forces are labelled Normal, Air resistance, Friction, Gravity, and Applied.

By using realistic values in Equation 9.3 for the cyclist's mass and the applied and resistive forces, their acceleration can be estimated. The higher the applied force, and the lower the resistive forces, the faster the cyclist will accelerate.

Let's start with the resistive forces. From Chapter 3, we know that the rolling resistance force of a bike travelling on a flat straight is:

$$F_{rr} = \mu_r F_N \qquad \text{Equation 9.4}$$

where F_N is the normal force acting on the cyclist (i.e., mass × gravity), and μ_r is the dynamic rolling resistance coefficient. A "scrubbing correction factor" can also be added to the right-hand side of Equation 9.4 to account for any losses due to scrubbing of the tyres when riding on banks and corners (see ref. [255]). For simplicity this factor will be ignored. The rolling resistance coefficient can be assumed constant and around 0.002 to 0.008 [43, 255, 256]. Assuming a cyclist mass of ~100 kg and hence normal force of ~1,000 N, the rolling resistive force would be only ~2 to 8 N. As will be shown, this is much lower than the drag force acting on a cyclist at typical race speeds (e.g., 11~15 m/s, as per Figure 9.5).

As introduced in Chapter 4, the drag force acting on a cyclist is:

$$F_d = \frac{\rho A C_D u_c^{\ 2}}{2} \qquad \text{Equation 9.5}$$

where ρ is air density, A is their frontal area, C_D is their drag coefficient, and u_c is their velocity. Air density is ~1.2 kg/m³ for typical race conditions [45], although it is ~20% lower at the high-altitude Mexico City velodrome [251]. The frontal area of a cyclist can be obtained from a front-on photo. It depends on the rider size and position and is typically 0.2~0.7 m² [45, 251]. The drag coefficient of cyclists is typically measured in wind tunnel tests or estimated from computational fluid dynamics simulations [43, 44, 46]. It varies with cycling speed, but for simplicity can be assumed constant. Typical values for drag coefficient are 0.5~1.0 [45, 46] depending on factors such as riding position, equipment and clothing. For simplicity, the frontal area and drag coefficient are often combined to give "drag area" ($C_D A$), which is typically 0.2~0.5 m² [44, 46]. Using this

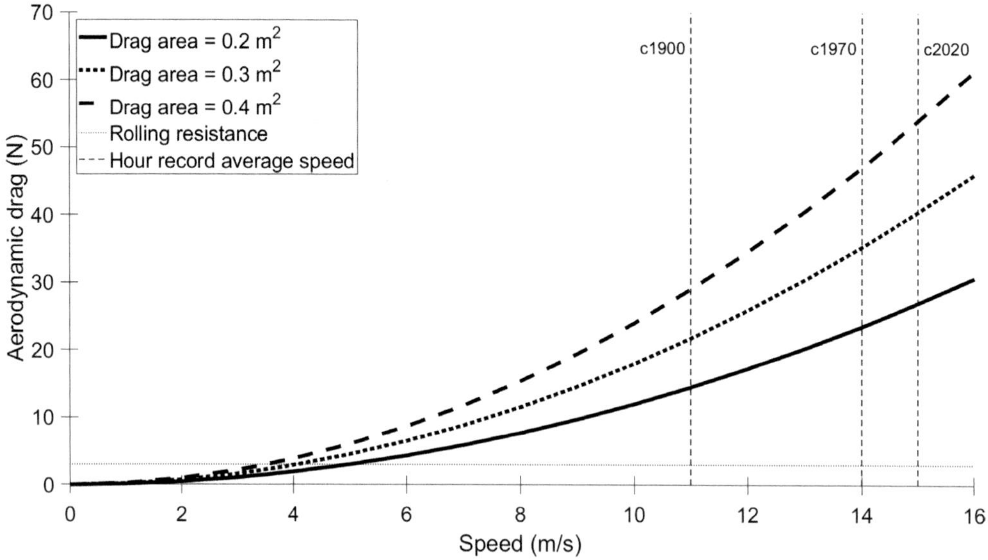

Figure 9.5 Graph showing how drag force increases with cycling speed for different values of drag area and soon becomes much larger than rolling resistance. The drag–speed relationships were calculated using Equation 9.5, assuming an air density of 1.2 kg/m³.

combination of drag area, it also becomes possible to compare the efficiency of riders with different sizes, positions, and speeds.

Figure 9.5 shows how drag force (Equation 9.5) increases with cycling velocity for different values of drag area and soon becomes much larger than the rolling resistance (i.e., $F_d \gg F_{rr}$). It is well known that ~90% of the resistive force acting on a cyclist at race speed is due to aerodynamic drag [43–46]. This means that Equation 9.3 can be simplified by removing rolling resistance to give Equation 9.6 without introducing much error:

$$a_c \approx \frac{(F_a - F_d)}{m_c} \qquad \text{Equation 9.6}$$

We now investigate the applied force from the cyclist. This force comes from the power they apply to the pedals (P_a):

$$F_a = \eta \frac{P_a}{u_c} \qquad \text{Equation 9.7}$$

where η is a factor to represent the bike's efficiency. Cycling power is typically measured with a power meter (e.g., [251, 255]). It varies between riders, and their maximum sustainable value decreases with cycling duration [43]. The maximum power a cyclist can sustain for an hour is 200~450 W [43, 251]. The factor representing the bike's efficiency encompasses that of the frame (η_f), bearings (η_b), and drive chain (η_d) (i.e., $\eta = \eta_f \eta_b \eta_d$). These individual efficiencies are high, and the total bike efficiency is ~94% (i.e., $\eta = 0.94$) [45, 255]. This means the efficiency can be ignored, simplifying Equation 9.7 to Equation 9.8, without introducing much error:

$$F_a \approx \frac{P_a}{u_c} \qquad \text{Equation 9.8}$$

Substituting Equation 9.5 and Equation 9.8 into Equation 9.6 gives Equation 9.9:

$$a_c \approx \frac{\left(\frac{P_a}{u_c}\right) - 0.5\rho A C_D u_c^2}{m_c} \qquad \text{Equation 9.9}$$

By using typical values (as outlined above) in Equation 9.9 and using small timesteps (i.e., time stepping technique), it is possible to estimate the acceleration and hence velocity of a cyclist throughout a race (see ref. [255]). However, after a brief period of initial acceleration, they will travel at an almost constant velocity for the remainder of the hour record attempt, provided the applied power remains constant. This means for most of the race their acceleration is zero and their applied force equals the drag force ($F_a \approx F_d$).

$$\frac{P_a}{u_c} \approx \frac{\rho A C_D u_c^2}{2} \qquad \text{Equation 9.10}$$

By rearranging Equation 9.10 to make the cyclist power the subject, we can see this increases with the cube of their velocity:

$$P_a \approx \frac{\rho A C_D u_c^{\,3}}{2} \qquad \text{Equation 9.11}$$

This means it becomes progressively much harder to cycle faster (Figure 9.6). Rearranging Equation 9.11 allows the average cyclist velocity ($\bar{u}_c$) to be estimated from their average power ($\bar{P}_a$) and typical values of air density and drag area:

$$\bar{u}_c \approx \sqrt[3]{\frac{2\bar{P}_a}{\rho A C_D}} \qquad \text{Equation 9.12}$$

This basic model can be used to gain a better understanding of the requirements for an hour record at different time periods (see Figure 9.2). It can also be used to gauge the effect of developments in bikes and associated equipment on performance. The vertical lines in Figure 9.6 show typical average speeds for the hour record at different periods. These correspond to circa 1900, 1970, and 2020, and respective distances of about 40, 50, and 54 km.

Circa 1900 corresponds to an average speed of ~11 m/s (~40 km/hr). Assuming a drag area for this early period of ~0.4 would require an average power of ~300 W. This seems reasonable for a high-level cyclist of that time. Moving on to circa 1970, the higher average speed of ~14 m/s (~50 km/hr) would require an average power >600 W for the same assumed drag area as circa 1900. Even for an elite cyclist this seems unrealistic. Reducing

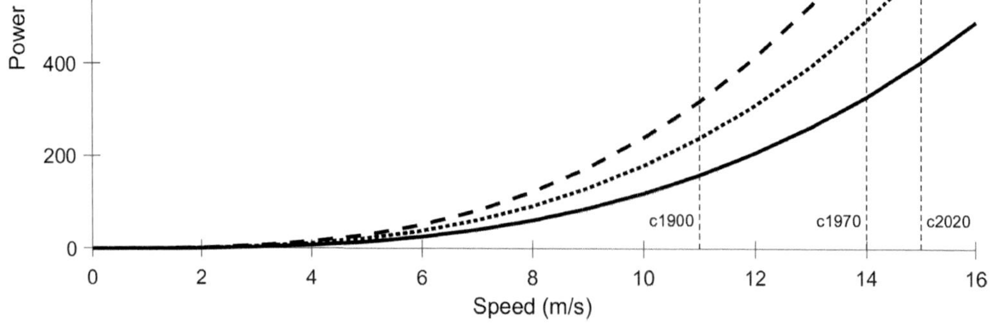

Figure 9.6 Graph to show how required power increases with cycling speed for different values of drag area. The power–speed relationships were calculated using Equation 9.11, assuming an air density of 1.2 kg/m³.

the drag area to ~0.3 to account for equipment developments reduces the average power to ~500 W, which seems more realistic, if perhaps a bit high. Finally consider circa 2020, with an average speed of ~15 m/s (~54 km/hr). The same assumed drag area (0.3) as circa 1970 would require an average power of ~600 W, which again seems unrealistic. The average power reduces to a more achievable ~400 W, if the drag area is reduced to 0.2 to account for modern equipment. This basic analysis further indicates the contributions advances in bike design, and particularly aerodynamic efficiency, have had on cycling performance. Indeed, a well-trained cycling enthusiast could now likely achieve the record distance of circa 1900 on a modern bike with a drag area of 0.2~0.3, by delivering an average power of 150~250 W.

Deeper analysis can be achieved with a more advanced version of the model that incorporates the efficiency of the bike and rolling resistance. Such a model could include the specific power profile and drag area of the cyclist, to accurately predict their one hour cycling distance (see ref. [255]). Such a model could also be applied to other cycling events. This would, however, be a time-consuming and high-resource endeavour. It would require extensive testing of athletes and their equipment to obtain the required power and aerodynamic inputs for the model. Further, applying such an approach to investigate historical performances would require further resource, access to rare and unusual bikes and equipment, and assumptions about rider characteristics. This seems impractical and prone to error. To overcome such issues, Haake [45] introduced the performance improvement index as a simple means of comparing historical performances between sports and gauging the effect of technology.

9.4.2 *Performance improvement index*

The performance improvement index will now be briefly summarised. This summary will start with the hour cycling record. From our previous analysis, we know that for most of the hour record, the applied force approximately equals the aerodynamic drag ($F_a \approx F_d$). The average drag force ($\bar{F}_d$) for a record attempt can be calculated by using the average velocity of the cyclist ($\bar{u}_c$) for their velocity (u_c) in Equation 9.5. Replacing this average velocity with the ratio of distance (s) and time (t) (see Chapter 3) then gives:

$$\bar{F}_d = \frac{\rho A C_D s^2}{2t^2} \qquad \text{Equation 9.13}$$

Work done ($Work$) is the product of force and distance (s). Therefore, the work done in overcoming the drag force is:

$$Work = \bar{F}_d s = \frac{\rho A C_D s^3}{2t^2} \qquad \text{Equation 9.14}$$

Haake's performance improvement index [45] compares the work done in two races:

$$Index = \frac{Work_2}{Work_1} = \frac{\left(\dfrac{\rho A C_D s^3}{2t^2}\right)_2}{\left(\dfrac{\rho A C_D s^3}{2t^2}\right)_1} = \frac{\left(\dfrac{\rho A C_D s^3}{t^2}\right)_2}{\left(\dfrac{\rho A C_D s^3}{t^2}\right)_1} \qquad \text{Equation 9.15}$$

As the time in the cycling hour record is fixed, the times for both races are the same and Equation 9.15 simplifies to:

$$Index = \frac{\left(\rho A C_D s^3\right)_2}{\left(\rho A C_D s^3\right)_1} \qquad \text{Equation 9.16}$$

If the air density and drag area for each race are known, these can be inserted into Equation 9.16 to estimate their influence. In the simplest case, these can be assumed to be the same between both races, to give the overall improvement performance index as a percentage.

$$Index(\%) = \left[\left(\frac{s_2}{s_1}\right)^3 - 1\right] \times 100 \qquad \text{Equation 9.17}$$

Figure 9.7 shows how the performance improvement index for the cycling hour record has progressed since the 1890s. This performance improvement index was calculated by applying Equation 9.17 to the data presented in Figure 9.2. The first distance (s_1) was taken as the earliest record in Figure 9.2 (~35 km in 1893). Values for the second distance (s_2) were taken as the corresponding value for each subsequent record. This means the performance improvement index shows the percentage improvement

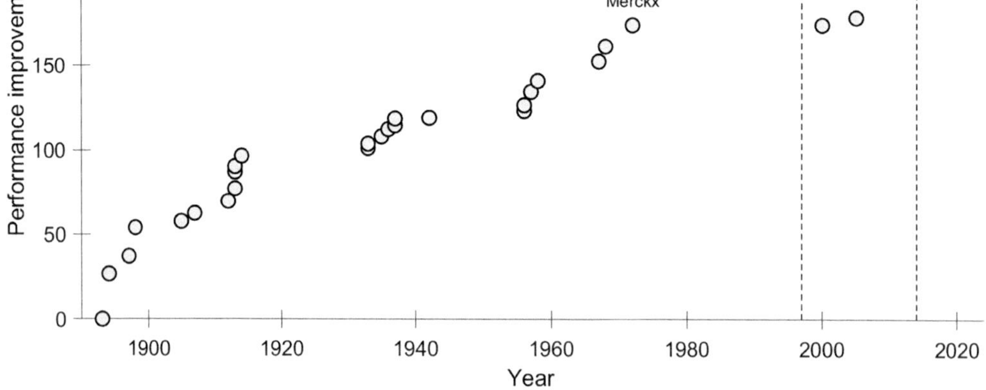

Figure 9.7 Graph showing how the performance improvement index for the cycling hour record has progressed by ~300% since the 1890s to present. The performance improvement index was calculated by applying Equation 9.17 to the data presented in Figure 9.2.

for each record relative to the distance of ~35 km in 1893. Haake [45] used a slightly later and hence longer first distance (~38 km in 1894), meaning his percentage improvements were lower than those shown here (e.g., 221% vs. 306% in 1996 and 120% vs. 187% in 2005).

Figure 9.7 shows a progression in the performance improvement index of 316% from 1893 to 2022. The improvement was ~174% from 1893 to the record in 1974, which can be considered as the last one before the uptake of aerodynamic drag-reducing features leading to the first rule change (see Figure 9.3). This means the remaining improvement from 1974 to 2022 was 142% (316–174%). As both rule changes were related to aerodynamic developments, it could be concluded that these contributed ~140% to the overall performance improvement index since 1893. By assuming the same air density for all records, it is also possible to estimate the effect of reductions in aerodynamic drag from equipment developments by simplifying Equation 9.16 to give Equation 9.18:

$$Index(\%) = \left[\frac{\left(AC_D s^3\right)_2}{\left(AC_D s^3\right)_1} - 1\right] \times 100 \qquad \text{Equation 9.18}$$

and using appropriate values for each drag area. The first step is to estimate drag area values for the records of interest. For simplicity, let's assume drag areas of 0.4, 0.3, and 0.2 m^2 for the respective records in 1893 (35 km, 0% improvement), 1972 (49 km, 174% improvement), and 2022 (57 km, 316% improvement). Comparing 1972 to 1893 gives:

$$Index(\%) = \left[\left(\frac{0.3\times 49^3}{0.4\times 35^3}\right) - 1\right] \times 100 = \left[\left(\frac{36234}{17632}\right) - 1\right] \times 100 = 106\% \qquad \text{Equation 9.19}$$

Note: The values shown in this equation are rounded.

This indicates that 106% of the improvement was from all other factors related to athlete performance, such as training and diet, with the remaining 69% (174–106%) from developments that reduced aerodynamic drag. Comparing 2022 to 1893 gives:

$$Index = \left[\left(\frac{0.2\times 57^3}{0.4\times 35^3}\right) - 1\right] \times 100 = \left[\left(\frac{36635}{17632}\right) - 1\right] \times 100 = 108\% \qquad \text{Equation 9.20}$$

Note: The values shown in this equation are rounded.

This indicates that 108% of the improvement was from all other factors related to athlete performance, with the remaining 208% (316–108%) from aerodynamic developments. Comparing improvements from aerodynamic developments from 1972 to 2022 gives 139% (208–69%), which is similar to the previous estimate. These basic estimates were clearly dependent on the rather arbitrary drag area values selected, and more representative values for these would give more accurate predictions.

We will now show how the performance improvement index can be applied to other sports, including running and pole vault. As running records are contested over a set distance, Equation 9.21 applies:

$$Index = \frac{\left(\frac{\rho A C_D s^3}{2t^2}\right)_2}{\left(\frac{\rho A C_D s^3}{2t^2}\right)_1} = \frac{\left(\frac{\rho A C_D}{t^2}\right)_2}{\left(\frac{\rho A C_D}{t^2}\right)_1} = \frac{(\rho A C_D)_2 t_1^2}{(\rho A C_D)_1 t_2^2} \qquad \text{Equation 9.21}$$

If air density is again assumed to be constant between records, Equation 9.21 simplifies to:

$$Index = \frac{(A C_D)_2 t_1^2}{(A C_D)_1 t_2^2} \qquad \text{Equation 9.22}$$

As with the hour cycling record, Equation 9.22 can be applied to historical running records to estimate the performance improvement. Again, the drag area can be set as constant or adjusted to account for improved aerodynamics due to tighter fitting clothes. Haake [45] reported a performance improvement index of 24% in the 100 m sprint from 1900 to 2008, estimating 4% of which came from aerodynamic improvements. The estimated contribution of aerodynamic improvements was clearly much lower for running than cycling, as expected.

For pole vault, where height is the performance metric, and work is the potential energy (see Chapter 3), Equation 9.23 applies:

$$Index = \frac{Work_2}{Work_1} = \frac{(mgh)_2}{(mgh)_1} = \frac{(mh)_2}{(mh)_1} \qquad \text{Equation 9.23}$$

Equation 9.23 simplifies to Equation 9.24 if athlete mass is assumed constant:

$$Index(\%) = \left[\left(\frac{h_2}{h_1}\right) - 1\right] \times 100 \qquad \text{Equation 9.24}$$

Figure 9.8 shows how the performance improvement index for the pole vault world record has progressed since the 1840s. This performance improvement index was calculated by applying Equation 9.24 to the data presented in Figure 1.5. The first height (h_1) was taken as the earliest record in Figure 1.5 (~3 m in 1849). As with distance in the hour cycling record, values for the second height (h_2) were taken as the corresponding value for each subsequent record. Again, the percentage improvement depends on the first height selected; hence our data differs from that of Haake [45] who started with 1900.

Figure 9.8 shows a progression in the performance improvement index for the pole vault of almost 100% since the 1840s to present. Clear improvements can be seen with the introduction of bamboo and then composite poles (see Chapter 4). Unlike their

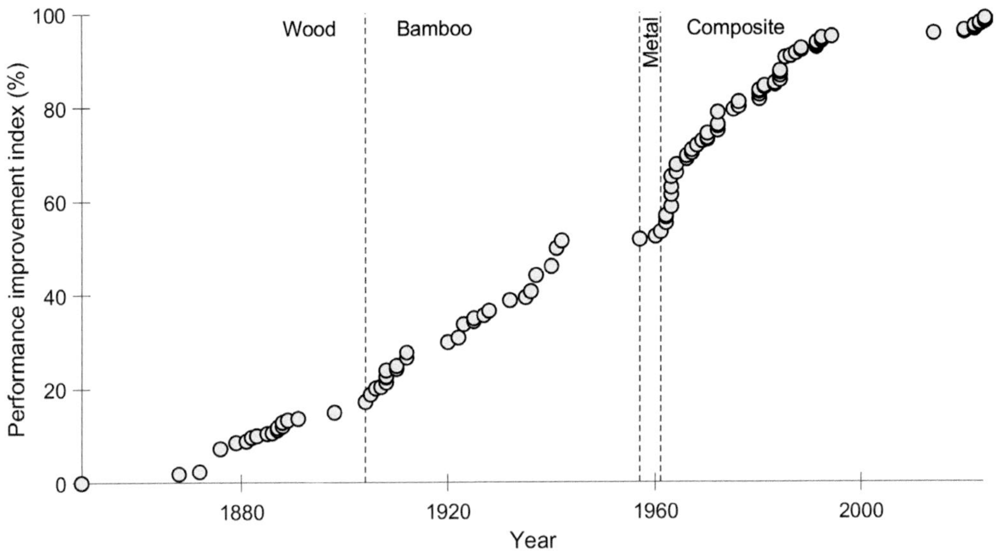

Figure 9.8 Graph showing how the performance improvement index for the pole vault has progressed by ~100% since the 1840s to present. The performance improvement index was calculated by applying Equation 9.24 to the data presented in Figure 1.5 for pole vault world record height.

wooden predecessors, bamboo poles were naturally hollow, which made them lighter and allowed the athlete to run faster. Composites bought greater design freedom and higher failure stress, allowing light, strong, and flexible poles that are good at storing and returning strain energy. The overall performance improvement index for pole vault sits between the 100 m sprint and hour cycling record, as might be expected given the relative amount of technology in each sport, and the contribution of aerodynamics to cycling.

While the performance index allows comparison between sports, it is important to remember that the specific values estimated depend on the first one used as the baseline. World records are also inherently sporadic, and there can be long gaps between them. A more complete picture can be obtained using a certain number (e.g., 25) of the best performances each year [253, 254], although obtaining these may be challenging.

Modelling techniques like those presented here can be developed, adapted, and applied to various sports and sporting scenarios (e.g., [1, 45, 249, 252–258]). This can, for example, include modelling the flight of a ball, such as those used for golf, tennis, and football, to predict how historical developments in design have influenced performance (e.g., number and shape of panels or dimples or both). Impacts between a ball and implement, such as a bat or racket, or sports surface can also be modelled, and even combined with flight models to give a holistic picture of the effect of design changes [249, 257]. Such models will ultimately depend on their complexity and accuracy, with some requiring more underpinning data, such as aerodynamic or material properties, than others. The sports engineer must therefore carefully consider their available resources. The best approach to gauging the effect of technology on sport is often to combine the various approaches presented in this chapter to give a holistic overview.

9.5 Large datasets and their role in assessing the effect of technology on sport

Another approach to gauging the effect of technology on sport is to analyse large datasets from sporting events. This typically means investigating whether performance benefits expected from new or improved equipment transfer to competitions. The expected benefits typically come from established research methods, such as predictive models (as already described in this chapter) or laboratory-based experiments (see Chapter 8). The widespread availability of large datasets from sporting events has come from technological developments and social trends, such as better computing, automated timing systems, ball and athlete tracking tools, and the rise of public data repositories and social fitness networks. This means analysis on large datasets typically focuses on recent developments in sports technology rather than long-standing historical performances (e.g., records spanning >100 years), as shown previously. An advantage of using large datasets is overcoming the inherent uncertainty and limitations associated with typically small-scale experiments. This subsection shows examples of how large datasets can be used, including for distance running and ball flight.

9.5.1 Distance running shoes

Distance running shoes have developed considerably, helping to improve running performances. The modern running shoe, with its polymeric closed cell foam (e.g., Ethylene Vinyl Acetate [EVA] foam) midsole, was introduced in the 1960s. A period of rapid development in running shoes started circa 2016 with the emergence of so-called "super shoes", or advanced footwear technology (see Figure 9.1b). Such footwear is lightweight with a curved stiff plate embedded in a thick foam midsole. This foam is compliant and resilient. Advanced footwear technology has attracted scientific and media attention as it seems to allow distance runners to perform better and break records, particularly in long-distance events.

There are many articles on advanced footwear technology (e.g., [53–56, 154, 258–266]). The consensus is that this footwear can improve running performance, including in the marathon. Due to concerns over the performance benefits of advanced footwear technology, World Athletics revised their rules in 2020, with further updates as knowledge and the general situation developed. These revisions included limiting midsole thickness and only allowing one stiff plate. Advanced footwear technology is now used here to showcase how sports engineers can use large datasets to gauge the effect of technology on sporting performance.

Early studies on advanced footwear technology focused on laboratory-based testing with runners (e.g., [53, 55, 56, 154]). Many of these compared measures of running economy on a treadmill when wearing different footwear. Running economy is essentially a measure of oxygen consumption at a constant sub-maximal speed. Improved economy means greater efficiency, allowing a runner to maintain their speed with less effort, and hence for longer, as is crucial for the marathon. Studies agree that advanced footwear technology can improve running economy by a few percent, although each runner responds differently, and the biomechanical reasons remain unclear. A key question is whether this laboratory-observed improvement translates to real-world marathons, allowing runners to maintain a higher speed and improve their overall performance.

Studies have used datasets of recent marathon times to observe the effect of advanced footwear technology on performance [54–56, 258, 259, 263, 266]. This approach differs from the examples earlier in the chapter, which used sparce and sporadic data over long periods (i.e., historical world records). The investigation period usually spans from just before the emergence of advanced footwear technology (circa 2016) to the date of the study. One approach is simply to observe how marathon performances over that period have changed, such as comparing the mean of the top 100 times each year, the number below a certain sub-elite time threshold (i.e. strong performances) and investigating whether the second half of the race was run faster than the first [54, 262]. Another approach has been to specifically compare performances with and without advanced footwear technology. The footwear used was either reported by the runner or identified by the researchers in publicly available photographs.

Such retrospective observational studies allow inclusion of a broader range of runners. These runners would have likely had time to become familiar with any changes in their footwear. These studies do, however, require careful design and robust statistical analysis to control for confounding factors, such as environmental conditions, race location, training, and the Covid-19 pandemic. The findings of large dataset studies generally agree with their laboratory-based counterparts, concluding that advanced footwear technology can, to a lesser degree, improve marathon performance. Findings also suggest that females benefit more from advanced footwear technology than males, who tend to be the focus of traditional laboratory-based studies [221].

9.5.2 *Sports ball flight*

The next case study will summarise how big data can be applied to investigate parameters that influence the flight of sports balls. The turn of the 21st century saw the introduction of camera-based ball tracking systems to sports like cricket, tennis, golf, and baseball [14]. Specialist software tracks the position of the ball in three dimensions, using interpolation techniques to calculate the flight path and impact locations. The established approach for investigating sports ball aerodynamics is to use a wind tunnel [125] (see Chapter 4). Another approach is to analyse the ball in free flight, which involves projecting it and measuring its path [10, 267, 268]. A trajectory model is then fitted to the flight path to obtain the ball's aerodynamic coefficients.

Using ball flight data obtained from tracking systems during competitions allows more trajectories to be analysed than during laboratory-based experiments. Choppin et al. [248] used tracking data for >9,000 tennis serves, spanning five years, to analyse the effect of ball wear on aerodynamics. By fitting trajectory models to the measured flight path, they showed that used balls had higher drag coefficients, as found with wind tunnel studies. This study showcases one of the many ways tracking technologies can be used to predict how changes to equipment can influence sport, which could include adaptions to the design of the ball, striking implement, or playing surface. A barrier to this approach is access to the data, which is often owned by the competition host or governing body, and not publicly available.

The case studies presented have highlighted how a big data approach can complement the findings of controlled laboratory-based studies. Both approaches have benefits, and often the best option for sports engineers is to combine different methods, such as reviewing historical records, artefacts, images, and videos [30, 269, 270], and undertaking modelling, laboratory- and field-based studies, and big data analysis. Such a holistic systems approach can give a clearer picture of the effect of past, current, and emerging technology on sporting performance. This can help sporting goods brands to develop products that benefit more people while allowing governing bodies to make informed regulatory decisions. It can also help governing bodies to "horizon scan", and ensure their regulations are "future-proof", e.g., allowing them to prepare for any performance enhancements that may come from a widespread uptake of metamaterials [15, 22, 23] within sporting goods.

9.6 Ethical considerations of technology in sport

There are ethical considerations, and different opinions, on the use of technology to enhance sporting performance [42, 64]. Such use of technology is often called "technology doping", with some considering it akin to using performance-enhancing drugs. Debate often centres around issues of fairness and equality, the distinction between human and technological performance, and the effect on fan experience and the integrity of sport. Some athletes and nations may not have access to certain technologies or lack the funds to support sports engineers to develop them, rendering them less competitive. There is also a distinction between technological officiating aids, such as the video assistant referee (VAR) in football [271] and its impact on fairness and the sport's integrity, and technology used to help disabled athletes to engage and compete in sport.

Regardless of opinions and controversies, it is vital for research investigating the effect of technology on sport to continue. Such work will inform the design of better sporting goods and help governing bodies make informed regulatory decisions about technology that improve rather than undermine or devalue the experience of athletes and fans. A key consideration is to avoid situations where new technologies offer such a benefit that the outcome of a competition is seen to be solely about which athlete has the best equipment. In these situations, technology could be deemed to be helping certain athletes to circumvent the sporting test or challenge unfairly, and perhaps even considered to be changing the inherent nature of the sport. It is also important to understand how technology can facilitate and enhance disability sport, while maintaining fairness between people with different impairments. We should also not forget the role technology can have on facilitating sport amongst a diverse populace, while increasing safety [152].

9.7 Implications for equipment development, regulation, and training

Studying the effect of technology on sport can have important implications for equipment development, regulation, and the training of athletes. Enhanced knowledge allows sports engineers to focus on worthwhile equipment developments, including those that make sport more accessible and widen participation. This is particularly important for recreational activities, like snow and roller sports, mountain biking, and jogging, which offer physical and mental health benefits. Regarding competitive sport, better knowledge of the effect of technology on performance and safety can inform athlete training methods and equipment selection and setup. This can lead to better performances and new records and enhance the spectacle of sport.

9.8 Chapter summary

This chapter of the textbook has outlined the importance of gauging the effect of technology on sport, while showcasing methods and considerations for doing so. The various methods presented include reviewing historical records and photographs of athletes and their equipment, predictive modelling, and analysing datasets from sporting events. Each of these approaches has advantages and limitations, and sports engineers should adopt a systems approach, using all of them to gather evidence and gain a holistic overview of the effect of technology on sport. They must remember the reasons for doing so, and the broad benefits, potential pitfalls, and ethical considerations of using technology to enhance sporting performance. A key factor is to ensure appropriate regulation of sports technology, which can come from research and horizon scanning to identify emerging developments. The next chapter of the textbook covers this topic by summarising current and emerging trends in sports engineering.

9.9 Suggested further reading

The following suggestions for further reading go deeper into the topics introduced in this chapter:

- "A comparison of the energetic cost of running in marathon racing shoes" [154]
- "Comparing cycling world hour records, 1967–1996: modeling with empirical data" [251]
- "Cyclist aerodynamics through time: better, faster, stronger" [44]
- "Does advanced footwear technology improve track and road racing performance? An explorative analysis based on the 100 best yearly performances in the world between 2010 and 2022" [258]
- "Ten questions in sports engineering: supershoe use in distance running" [56]
- "The controversy of sports technology: a systematic review" [42]
- "The effect of ball wear on ball aerodynamics: an investigation using hawk-eye data" [248]
- "The impact of technology on sporting performance in Olympic sports" [45]

Notes

1 https://www.randa.org/en/equipment
2 https://www.usga.org/content/usga/home-page/equipment-standards.html
3 https://www.itftennis.com/en/about-us/tennis-tech/

10 Current trends and future challenges in Sports Engineering and Technology

This final chapter discusses some current trends, challenges, and problems facing sports engineers, drawing on themes introduced throughout the book. The topics covered are:

1. Smart sporting goods
2. Mass market customisation of sporting goods
3. Ensuring access to sport, including accounting for the specific needs of females and disabled athletes and participants
4. Environmentally sustainable manufacturing of sporting goods
5. Injury risk reduction, including traumatic brain injury
6. The increasing importance of big data and artificial intelligence in understanding the factors contributing to sporting performance and injury risk

The importance of embedding environmental sustainability into the sporting goods industry is reiterated. Many of these topics reflect those in the recent IMechE policy report on sports engineering [15].

10.1 Smart sporting goods

As discussed throughout the book, smart sports equipment and clothing is an emerging and expanding topic [18]. Such smart items have potential to react and adapt to the movement patterns and needs of the user. This often involves embedding electronics within sporting goods, to sense, monitor, and drive change. Imagine outdoor clothing with temperature sensors for monitoring thermal comfort. Such clothing could react accordingly, like offering feedback and guidance to the wearer, whilst actively changing its thermal properties.

Wearable items like fitness trackers and activity monitors have become common sporting accessories, particularly for runners (e.g., Figure 10.1a). We may soon see closer links between such items and smart sports equipment and clothing. Imagine smart shoes that adapt their properties based on metrics such as heart rate, temperature, and distance travelled, to maximise performance and reduce injury risk. There may be a move towards a systems approach for smart sporting goods. This could involve items that work together utilising concepts like cloud computing, the internet of things, digital twin, and virtual reality (e.g., Figure 10.1b). Such an approach could have broad implications across many sport- and exercise-related topics. These could include activity monitoring, health predictions, consumer trends and equipment use, and user and fan experience and enjoyment, all while facilitating remote competition, training, and socialising.

DOI: 10.4324/9781003411086-10

(a)

(b)

Figure 10.1 Images showing concepts of smart sporting goods, (a) activity monitoring smart watch with fitness metrics overlaid, (b) a virtual reality enhanced online workout, and (c) an integrated collection of wearable technology.

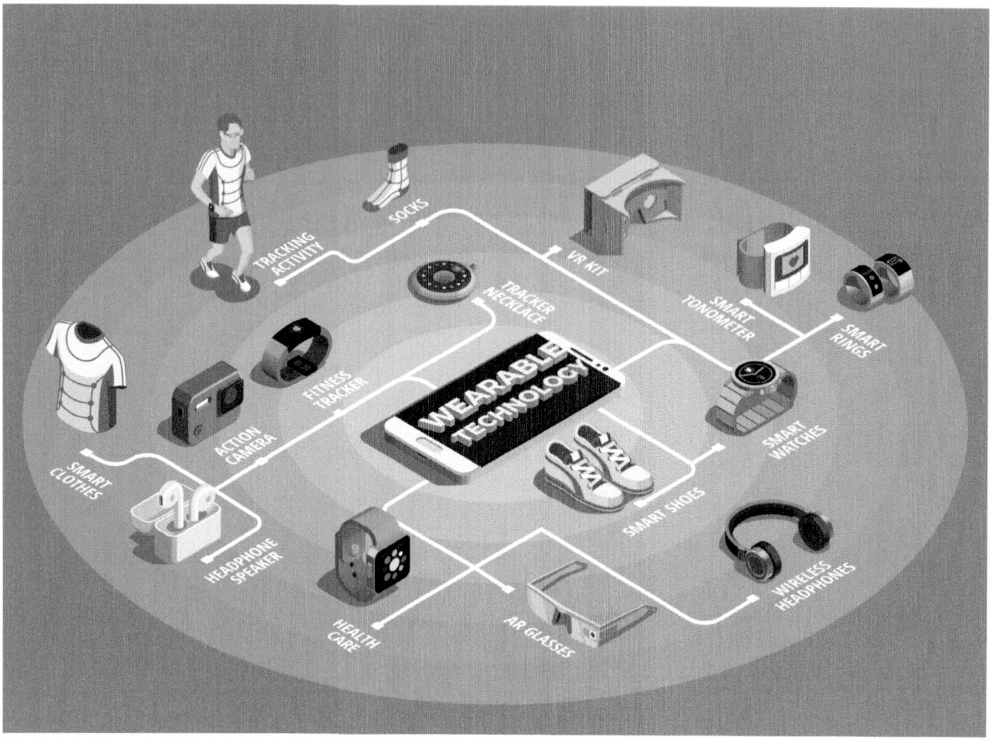

(c)

Figure 10.1 (Continued)

There are clear challenges associated with realising the potential of smart sporting goods. These include technical challenges and those related to the feasibility, cost, and sustainability of mass producing such items [15]. These challenges could be overcome by adopting a multifaceted collaborative approach. This could involve bringing together people with expertise in topics like metamaterials, wearable technologies, sports science, healthcare, computing, machine learning, and sustainability. Embedding electronics within sporting goods bring sustainability challenges, spanning the environmental impact of raw materials, energy consumption during use and end-of-life disposal. Sports engineers developing smart sporting goods will have to embrace new and emerging sustainability regulations, such as the EU Batteries Regulation (2023/1542).

10.2 Mass market customisation of sporting goods

Current sporting goods tend to be mass-produced for consumers to benefit from economies of scale (e.g., Figure 10.2a and b). Future consumers are likely to become increasingly demanding, expecting more from their products and a customised or bespoke approach. Customising products on a large scale, while keeping end costs reasonable and minimising environmental impact, is clearly challenging. Sporting goods brands will

(a)

(b)

Figure 10.2 Photos illustrating (a) mass production of bike frames, (b) mass production of sports shoes, and (c) the concept of instore gait analysis.

(c)

Figure 10.2 (Continued)

need to embrace new and emerging technologies and concepts if they want to address the needs of an increasingly demanding customer base. Some possible trends are now briefly discussed.

As noted in the previous subsection (10.1), we may see an emergence of smart sports products that adapt, and hence customise themselves, to the user. Such products could help negate the need for sporting goods brands to increasingly diversify their ranges to cater for different customers. We may also see a move towards "smart factories" and connected online ordering systems. Such an approach could allow customers to select from various options, like the colour or material of parts, to give the impression of a customised product. We could also see a move towards modular products, e.g., with customers choosing from a range of parts, which they assemble to form a customised product, somewhat like building a custom bike. We may also see a move towards a more immersive and customised shopping experience, with instore technologies used to help match customers to products (e.g., Figure 10.2c). This could, for example, involve using tools like body scanners and wearable sensors to find the best-fitting product. A key challenge for sports engineers would be to bring such an approach to the customer, while keeping costs desirable.

10.3 Ensuring access to sport

Sport must be accessible to all. This extends to the devices used by athletes and participants [15]. These devices must cater for the diverse needs of the user. Figure 10.3

from the IMechE policy report on sports engineering [15] highlights potential barriers to sports participation. Sports engineers must consider these barriers when developing sporting goods and associated technologies. This often means adopting an inclusive design approach. Many of the topics of this chapter, such as adaptive sports equipment and remote training and competition tools, could help overcome barriers to sports participation. Sports engineers must also include a diverse range of participants in their studies, as relevant to the research question [221]. This is to ensure the study results, and any associated sports products that may come directly or indirectly from it, cater for the needs of the intended population. Sports engineers aiming to develop accessible sporting goods will likely face various challenges and constraints, including budgetary, with many brands currently focusing on the core customer base.

10.4 Environmentally sustainable manufacturing of sporting goods

Concern over climate change, and the environment in general, necessitates the need for a more sustainable sporting goods industry [15]. Huge volumes of sporting goods end their life as waste (e.g., Figure 10.4a), and this practice must end. There are various aspects that contribute to the environmental impact of the sporting goods industry. Some new and emerging trends and measures for addressing these are summarised here.

Consumers, and sports fans, are becoming more aware of the environmental impact of sporting goods. As such, we expect them to help drive change. For example, customers may start demanding more insight into the environmental impact of their purchases, which could be verified by independently audited ecolabels (e.g., Figure 10.4b). Sporting governing bodies will also likely introduce their own environmental guidelines and requirements for sporting goods (e.g., [272]). Such labels and regulations could span raw materials and production processes, product durability and lifespan, and what happens to items no longer being used. Governmental regulations will also drive change. For example, the new EU Regulations concerning microplastics (Commission Regulation (EU) 2023/2055) have broad implications for artificial turf and related research [189–191]. Sporting goods brands will need to embrace new regulations and standards as they continue to develop products for an increasingly environmentally conscious customer.

We expect sporting goods brands to move towards more environmentally circular business models. We may see reduced reliance on certain materials. This could include moving away from synthetic materials to those that are natural, although this brings challenges. Manufacturers are also looking to improve production efficiency, which includes reducing waste and reusing scrap material. Building on the themes of this chapter, this could mean a move towards highly efficient smart factories, located closer to target customers. Such customers are also likely to expect their purchases to be more durable and last longer.

We could also see users changing their habits to reduce waste and environmental impact. For example, there could be a shift towards reselling and rental (e.g., Figure 10.4c and d), and long-term borrowing or subscription schemes. Such schemes could see rental shops and subscribers exchanging old items for new ones, so they can be reconditioned or recycled and fed back into circulation. To address such changes and environmental challenges, manufacturers are exploring ways of easing recycling of used products. For example, designing for disassembly, which can include using fewer materials and parts and avoiding complex joining and bonding techniques. Metamaterials offer potential in this space, as different properties can be obtained from fewer materials via a change in

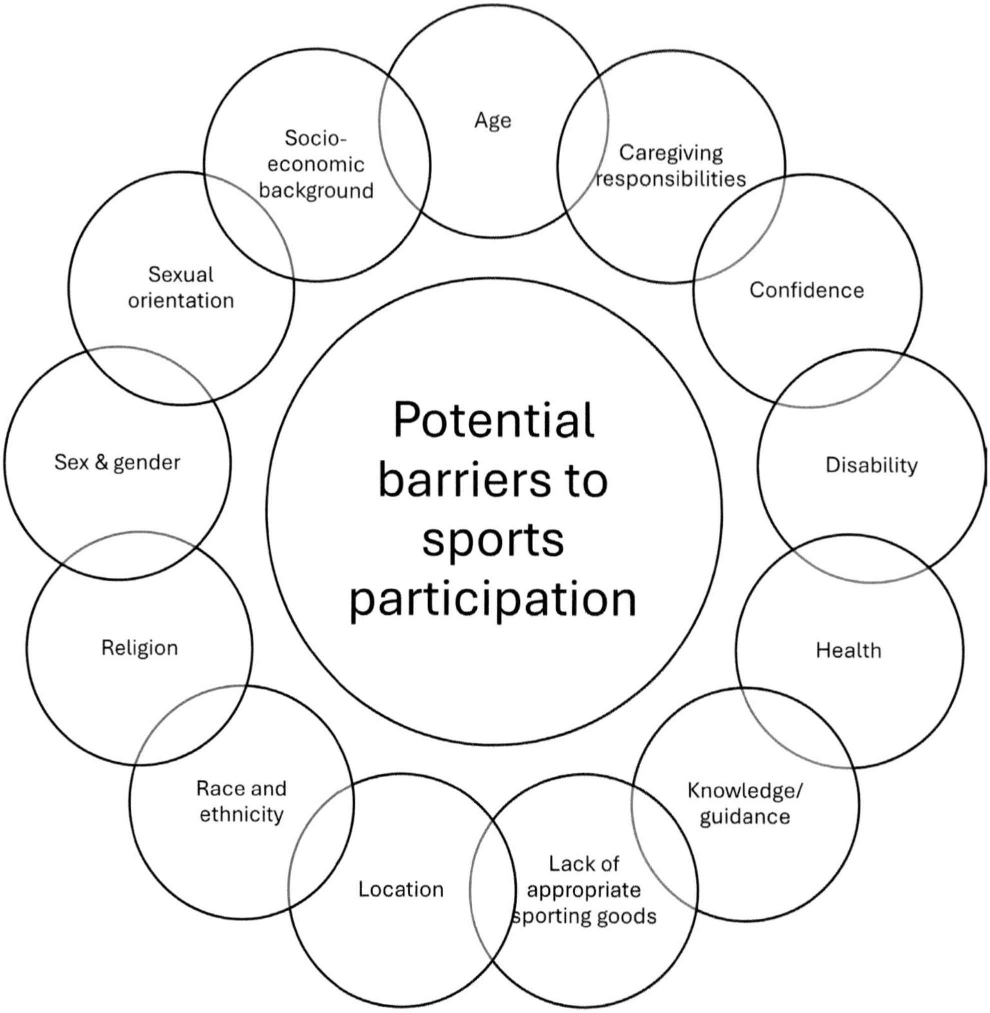

Figure 10.3 Illustration showing potential barriers to sports participation, based on image from ref. [15].

the geometric structure. Indeed, the topic of environmental sustainability presents many research challenges and opportunities for sports engineers.

10.5 Injury risk reduction

Injury risk is an inherent feature of sport and hence will remain a challenge and research opportunity for sports engineers. Athletes will remain exposed to the different categories of injuries, including overuse, acute, and traumatic/fatal. With increasingly busy event and competition schedules combined with increasing demands from fans for a spectacle, injury risk for elite athletes could increase. Amateurs are also at risk of injury, particularly if they are underprepared, using low-quality or ill-fitting equipment,

(a)

(b)

Figure 10.4 Photos showing examples related to the effect of sporting goods on the environment, (a) abandoned ski boots, (b) an ecolabel on clothing, which should ideally be independently audited, and sporting goods rental services, including (c) bikes and (d) ski boots.

(c)

(d)

Figure 10.4 (Continued)

or participating in poorly designed or unsuitable venues. For example, a beginner mountain biker riding an unsuitable bike on an advanced-grade downhill track could be at high risk of injury. Changing trends, new sports, and climate change could all influence injury risk. For example, reduced snow coverage in snow sports resorts could increase injury risk.

By working collaboratively with experts from different disciplines and taking a multi-faceted approach, sports engineers will be able to reduce sporting injury risk. A good example of this approach is that of the International Society for Snowsports Safety (ISSS).[1] Key aspects of reducing injury risk will include raising the quality and performance of sporting goods, including protective devices, and venues (e.g., snow sports terrain parks and bike tracks) via the development of standards, regulations, and unified guidelines (e.g., [152]). Other advances will likely include the development of technology-based training tools (e.g., wearable devices [18] and virtual reality [273–275]), signage, reporting, and diagnosis. One area where further research and development are expected is traumatic brain injury and concussion risk. We will likely see the development of new standards and helmet technologies, including those featuring metamaterials [22]. We will also likely see a move away from an androcentric (male-centred) approach to the design of sporting goods and venues to one that is more inclusive and hence reduces injury risk for females.

10.6 Big data and artificial intelligence

Big data and artificial intelligence will continue to impact all aspects of society, including sport. Sports engineers will continue to collect increasingly large volumes of data and use artificial intelligence to address research questions. These questions will span topics related to sporting performance and injury risk, as well as fan experience and consumer trends and preferences. Such monitoring will then likely feed back into the design of sporting goods. Artificial intelligence can also be applied to the design process itself, such as by proposing diverse candidate designs based on user specifications. Such an approach could be particularly useful in the design of complex metamaterial structures [22]. The increasing need for data builds on the themes of this chapter, particularly smart sporting goods. We will also likely see an increase in data collection during competitions and events. Artificial intelligence could also lead to new sports, bringing further opportunities for sports engineers. Sports engineers must take an ethical approach when collecting and handling data and using artificial intelligence.

After drafting this chapter, the "Future of Sports Engineering" was entered into the artificial intelligence writing tool ChatGPT,[2] which gave the following response, shown in Box 10.1.

Box 10.1

Start of ChatGPT response to "Future of Sports Engineering"...

> *The future of sports engineering is poised for significant innovation, driven by advancements in technology, materials science, data analytics, and human*

performance optimization. Here are some key areas where sports engineering is expected to evolve:

1. *Wearable Technology & Smart Equipment*
   Wearables: *The development of advanced wearables (like smart clothing, sensors, and trackers) will continue to grow. These devices will monitor athletes' biometric data in real-time, such as heart rate, hydration levels, muscle strain, and fatigue. This data will help improve training, injury prevention, and recovery.*
   Smart Gear: *Sports equipment such as shoes, helmets, or gloves will become increasingly intelligent, integrating sensors to monitor an athlete's movements, impact forces, and biomechanical performance. For instance, smart football helmets might detect concussion-inducing impacts.*
2. *AI and Data Analytics*
   Performance Optimization: *AI-driven systems will analyze an athlete's performance in real-time to offer personalized coaching tips. Data analytics will also be used to track progress, optimize training regimens, and fine-tune tactics and techniques.*
   Injury Prediction & Prevention: *Machine learning algorithms will analyze vast datasets of injuries and player movements to predict and prevent potential injuries before they occur. This could reduce downtime and extend athletes' careers.*
   Game Strategy: *AI will revolutionize how teams prepare for matches. Using vast amounts of data, AI can simulate various game scenarios, offering strategies based on opponent analysis, player form, and game conditions.*
3. *Advanced Materials & Biomechanics*
   High-Performance Materials: *The development of lightweight, durable, and sustainable materials will lead to improvements in sports equipment, from racquets and bikes to clothing and footwear. For example, carbon fiber and graphene are expected to continue being used for their strength-to-weight ratio, allowing for faster and more efficient performance.*
   Biomechanics: *Advances in biomechanics will allow for better understanding and enhancing of athletic performance. Motion capture technology, combined with machine learning, will lead to more precise analysis of athletes' movements, enabling tailored coaching and injury prevention strategies.*
4. *Virtual Reality (VR) & Augmented Reality (AR)*
   Training: *VR will become an essential tool for training, allowing athletes to simulate real-game scenarios without physical limitations. VR can also be used for mental preparation, helping athletes visualize complex game strategies or pressure situations.*
   Fan Engagement: *AR and VR will transform how fans experience sports. Virtual stadiums and interactive game experiences will bring fans closer to the action, enabling them to watch games from different angles or participate in virtual training camps.*
5. *Robotics & Exoskeletons*
   Exoskeletons for Performance: *Exoskeletons, which assist with movement and support the body, could be used to enhance an athlete's strength, speed, and endurance, or help with rehabilitation and recovery post-injury.*

Robotic Training Partners: *Robots that can mimic human movements may become commonplace in training, acting as sparring partners or opponents for athletes, allowing for highly customized and controlled practice sessions.*

6. *Sustainability in Sports Equipment*
   Eco-Friendly Materials: *With growing environmental awareness, there will be a shift towards more sustainable and eco-friendly materials in the design and manufacture of sports equipment. Companies may start producing biodegradable or recyclable sports gear, reducing the ecological footprint of the industry.*
   Energy Harvesting: *Sports equipment like shoes or wearables might use kinetic energy to power sensors or other small electronic devices, reducing the reliance on batteries and promoting sustainability.*
7. *Sports Simulation & Virtual Training*
   AI-Driven Simulations: *The integration of advanced simulations and AI models will provide athletes with the ability to train in virtual environments that closely replicate real-world conditions, enabling a level of preparation that is difficult to achieve in traditional training.*
   Personalized Virtual Coaches: *Virtual training systems powered by AI and machine learning will act as personalized coaches, providing athletes with tailored workout programs, feedback, and advice based on their progress.*
8. *Fan Engagement and Interactive Sports*
   AI and Interactive Fan Experiences: *AI will offer fans personalized experiences, from smart stadiums that enhance in-person experiences to interactive digital platforms that allow remote fans to control camera angles, access in-depth stats, or participate in virtual sports events.*
   eSports and Virtual Sports: *The rise of eSports and virtual sports leagues will continue to expand the boundaries of sports engineering, blending physical and digital experiences and opening up new opportunities for fan engagement and competition.*
9. *Augmented Coaching & Decision-Making Tools*
   Instant Data for Coaches: *Coaches will have access to real-time data about an athlete's performance and health, which will aid in making informed decisions about player substitution, tactics, and overall game strategy. This could also be enhanced with AI predictions that suggest tactical adjustments mid-game.*
   Increased Use of Video Analysis: *High-definition video and motion tracking technologies will be further integrated into training and in-game decision-making, allowing coaches to break down each play in greater detail, making more strategic adjustments.*
10. *Human Performance & Bioengineering*
    Gene Editing and Biotechnology: *The potential for breakthroughs in bioengineering, such as gene editing or growth hormone regulation, could push human performance to new heights, albeit raising ethical questions.*
    Nutrition and Recovery Technologies: *Advances in nutrition, supplements, and recovery technologies (such as cryotherapy, hyperbaric chambers, or advanced physiotherapy tools) will help athletes recover faster and maintain peak performance throughout the season.*

11. *Conclusion*
    The future of sports engineering will be a fusion of cutting-edge technologies designed to enhance athletic performance, improve fan engagement, and address sustainability challenges. These advancements will not only make sports safer and more exciting but will also offer athletes an unprecedented level of support in their training, performance, and recovery. As technology continues to evolve, the intersection of AI, biomechanics, and data analytics will drive a new era in sports innovation.

. . . End of ChatGPT response.

10.7 Chapter summary

This chapter of the textbook has provided an overview of current trends and future challenges in Sports Engineering and Technology. These are intended to give an indication of how the field of sports engineering, and the sporting goods industry, may be developing. The purpose was not to try to predict the future, nor to cover every eventuality. Rather it was to provide an indication of the direction of travel to encourage thinking, discussion, and debate.

This textbook has provided a broad overview and introduction to the developing field of sports engineering. There is much to learn beyond what is covered here, and students are encouraged to continue reading and learning and engaging with the wider sporting engineering community. Some suggestions for further reading are provided below.

10.8 Suggested further reading

The following suggestions for further reading go deeper into the topics introduced in this chapter:

- *Routledge Handbook of Sports Technology and Engineering*, 1st and 2nd edition [10], [86]
- *Sports Technology: Technologies, Fields of Application, Sports Equipment and Materials for Sport* [276]
- *21st Century Sports* [246]

Notes

1 https://www.isss.ski/
2 ChatGPT.com

11 Other books on Sports Engineering and Technology

Table 11.1 Suggested further reading on sports engineering and related topics. These are not textbooks. They are arranged alphabetically by topic

Title	*Topic*
Physics of Baseball and Softball [92]	Baseball and softball
Chris Boardman: The Biography of the Modern Bike: The Ultimate History of Bike Design [47]	Bicycle design
Sports Biomechanics: Reducing Injury Risk and Improving Sports Performance, 2nd edition [218]	Biomechanics
Computational Intelligence in Sports [277]	Computational intelligence
Routledge Handbook of Sports Technology and Engineering, 1st and 2nd editions [10, 86]	Engineering and Technology
Routledge Handbook of Ergonomics in Sport and Exercise [278]	Ergonomics
Computational Fluid Dynamics for Sport Simulation [279]	Fluid dynamics
The Science of Footwear [280]	Footwear
Golf by the Numbers: How Stats, Math, and Physics Affect Your Game [281]	Golf
Materials in Sports Equipment, 1st and 2nd editions [133, 134]	Materials
21st Century Sports [246]	Perspective on how technology will change sport
Wearable Sensors in Sport: A Practical Guide to Usage and Implementation [231]	Sensors
Materials and Technology for Sportswear and Performance Apparel [167]	Sportswear
Gold Medal Physics: The Science of Sports [108]	Sports physics
The Science and Engineering of Sport Surfaces [247]	Sports surfaces
World Book of Swimming: From Science to Performance [282]	Swimming
Measurement of Human Locomotion [283]	Technology used in biomechanics
Technical Tennis: Racquets, Strings, Balls, Courts, Spin, and Bounce [284]	Tennis
The Physics and Technology of Tennis [285]	Tennis
The Engineering Approach to Winter Sports [286]	Winter sports
Gliding for Gold: The Physics of Winter Sports [109]	Winter sports

DOI: 10.4324/9781003411086-11

References

[1] S. Haake, *Advantage Play: Technologies that Changed Sporting History*. Edinburgh: Birlinn Ltd, 2018.

[2] A. Morisse, "Bottom for football shoes," US1797668A, 1929. Accessed: May 7, 2025. [Online]. Available: https://patents.google.com/patent/US1797668A/en?oq=US+1797668+A

[3] A. Dassler, "Socket with through threaded hole to accept studs under the sole of a sport shoe," DE1737733U, 1952. Accessed: May 7, 2025. [Online]. Available: https://patents.google.com/patent/DE1737733U/en?oq=DE1737733U

[4] L. Wacker, "Studs for football and similar sports shoes," DE 443 311, 1925. Accessed: Aug. 18, 2025. [Online]. Available: https://www.dpma.de/docs/dpma/veroeffentlichungen/de443311-1ludwigwacker.pdf

[5] A. Salot, "Football boots or the like with replaceable anti-skid studs," DE 815 761, 1949. Accessed: Aug. 18, 2025. [Online]. Available: https://www.dpma.de/docs/dpma/veroeffentlichungen/de815761b_alexandersalot.pdf

[6] V. Eigler, "Replaceable stud consisting of several leather discs fixed to a base plate with a metal pin and easily replaceable when worn," DE 530 454, 1930. Accessed: Aug. 18, 2025. [Online]. Available: https://www.dpma.de/docs/dpma/veroeffentlichungen/de530454-1venustuseigler.pdf

[7] Deutsches Patent- und Markenamt (DPMA), "Screw stud football shoe: whose invention." Accessed: Aug. 18, 2025. [Online]. Available: https://www.dpma.de/english/our_office/publications/milestones/inventionsthatmadehistory/screstudshoe/index.html

[8] National Operating Committee on Standards for Athletic Equipment (NOCSAE), "Standard performance specification for newly manufactured football helmets." Accessed: May 8, 2025. [Online]. Available: https://nocsae.org/nd002-standard-performance-specification-for-newly-manufactured-football-helmets/

[9] Lord Rayleigh, "On the irregular flight of a tennis ball," *Messenger of Mathematics*, vol. 7, pp. 14–16, 1877.

[10] T. Allen, A. Alderson, A. Godfrey, D. Knudson, J. Webster, and K. Seo, Eds, *Routledge Handbook of Sports Technology and Engineering*, 2nd ed., vol. 1. Manchester: Routledge, 2025.

[11] S. Haake, Ed., *The Engineering of Sport*. Sheffield: CRC Press, 1996, pp. 1–358.

[12] S. Haake, "Editorial," *Sports Engineering*, vol. 1, no. 1, 1998, doi: 10.1046/j.1460-2687.1998.01001.x.

[13] M. Caine, "Journal of Sports Engineering and Technology – an introduction from the editor," *Proc Inst Mech Eng P J Sport Eng Technol*, vol. 222, no. 1, 2008, doi: 10.1177/175433710822200101.

[14] IMechE, "Sports Engineering: an unfair advantage?" 2012. Accessed: Apr. 2, 2025. [Online]. Available: https://www.imeche.org/policy-and-press/reports/detail/sports-engineering-an-unfair-advantage

[15] T. Allen, C. Nettleford, and M. Rooney, "Sustainable, inclusive, innovative: the role of engineering in sport," 2023. [Online]. Available: https://www.imeche.org/policy-and-press/reports/detail/sustainable-inclusive-innovative-the-role-of-engineering-in-sport

[16] A. Coppel, K. Te Kani, L. Kent, and R. Shilston, "Engineering better sports venues: designing for athletes, spectators, and the environment," 2024. Accessed: May 1, 2025. [Online]. Available: https://www.imeche.org/policy-and-press/reports/detail/engineering-better-sports-venues

[17] K. Seo, "Special issue on JSME symposium: sports engineering and human dynamics," *Proc Inst Mech Eng P J Sport Eng Technol*, vol. 237, no. 2, 2023, doi: 10.1177/1754337123 1169447.

[18] T. Allen, J. Shepherd, J. Wood, D. Tyler, and O. Duncan, "Wearables for disabled and extreme sports," in *Digital Health*, A. Godfrey and S. Stuart, Eds, Newcastle Upon Tyne: Elsevier, 2021, ch. 16, pp. 253–273. [Online]. Available: https://linkinghub.elsevier.com/retrieve/pii/ B9780128189146000168

[19] A. Koptyug and B. Burkett, "Editorial for the special issue technology for disability sport," *Sports Engineering*, vol. 19, no. 3, 2016, doi: 10.1007/s12283-016-0211-0.

[20] M. Chamberlain *et al.*, "Development of a bicycle crank arm demonstrator via Industry 4.0 principles for sustainable and cost-effective manufacturing," *Sports Engineering*, vol. 26, no. 1, p. 2, Dec. 2023, doi: 10.1007/s12283-022-00394-1.

[21] S. Spurrier, T. Allen, and R. A. Grant, "Investigating foot morphology in rock climbing mammals: inspiration for biomimetic climbing shoes," *Biomimetics*, vol. 8, no. 1, p. 8, Dec. 2022, doi: 10.3390/biomimetics8010008.

[22] D. Haid *et al.*, "Mechanical metamaterials for sports helmets: structural mechanics, design optimisation, and performance," *Smart Mater Struct*, vol. 32, no. 11, p. 113001, Nov. 2023, doi: 10.1088/1361-665X/acfddf.

[23] O. Duncan *et al.*, "Review of auxetic materials for sports applications: expanding options in comfort and protection," *Applied Sciences*, vol. 8, no. 6, 2018, doi: 10.3390/app8060941.

[24] D. Plant, "Energy absorbing system," US20120021167A1, Dec. 29, 2009. Accessed: Mar. 31, 2025. [Online]. Available: https://patents.google.com/patent/US20120021167A1/en

[25] E. Bliven *et al.*, "Evaluation of a novel bicycle helmet concept in oblique impact testing," *Accid Anal Prev*, vol. 124, 2019, doi: 10.1016/j.aap.2018.12.017.

[26] T. Hoshizaki, A. M. Post, C. E. Zerpa, E. Legace, T. B. Hoshizaki, and M. D. Gilchrist, "Evaluation of two rotational helmet technologies to decrease peak rotational acceleration in cycling helmets," *Sci Rep*, vol. 12, no. 1, 2022, doi: 10.1038/s41598-022-11559-0.

[27] Government Office for Science, "RTA: Metamaterials – GOV.UK." Accessed: Mar. 13, 2025. [Online]. Available: https://www.gov.uk/government/publications/rapid-technology-assessment-metamaterials/rta-metamaterials

[28] L. Taraborrelli, S. Choppin, S. Haake, S. Mohr, and T. Allen, "Effect of materials and design on the bending stiffness of tennis rackets," *Eur J Phys*, vol. 42, no. 6, 2021, doi: 10.1088/1361-6404/ac1146.

[29] R. A. Grant, L. Taraborrelli, and T. Allen, "Morphometrics for sports mechanics: showcasing tennis racket shape diversity," *PLoS One*, vol. 17, no. 1 January, 2022, doi: 10.1371/ journal.pone.0263120.

[30] L. Taraborrelli *et al.*, "Materials have driven the historical development of the tennis racket," *Applied Sciences*, vol. 9, no. 20, Oct. 2019, doi: 10.3390/app9204352.

[31] H. Head, "Composite wood and metal ski having plastic running surface," 2,694,580, 1951. [Online]. Available: https://patents.google.com/patent/US2694580A/en

[32] H. Head, "Tennis racket," 3,999,756, 1976. [Online]. Available: https://patents.google.com/ patent/US3999756A/en

[33] S. Kuebler, *Book of Tennis Rackets from the Beginning in the 16th Century Until about 1990*. Singen: Kuebler GmbH, 2000.

[34] S. Kuebler, H. Lammer, and J. Kotze, "Materials and tennis rackets," in *Materials in Sports Equipment*," M. Jenkins, Ed., Cambridge: Woodhead Publishing, 2003, Vol. 1, pp. 222–248.

[35] R. C. Haines, M. E. Curtis, F. M. Mullaney, and G. Ramsden, "The design, development and manufacture of a new and unique tennis racket," *Proc Inst Mech Eng Part B*, vol. 197, 1983, doi: 10.1243/PIME_PROC_1983_197_043_02.

[36] S. Miller, "Modern tennis rackets, balls, and surfaces," *Br J Sports Med*, vol. 40, no. 5, p. 401, May 2006, doi: 10.1136/bjsm.2005.023283.

[37] W. Wingfield, "Improvement in game apparatus," US157259A, 1874. Accessed: May 2, 2025. [Online]. Available: https://patents.google.com/patent/US157259A/en

[38] U. Schoberer, "Powermeter for a crank drive," WO1989000401A1, 1988. [Online]. Available: https://patents.google.com/patent/WO1989000401A1/de

[39] F. Fairhurst and J. Cappaert, "Articles of clothing," US6446264B2, 1999. [Online]. Available: https://patents.google.com/patent/US6446264B2/en

[40] J. Connell *et al.*, "Stacked cushioning arrangement for sole structure," US20180213886A1. Accessed: Mar. 14, 2023. [Online]. Available: https://patents.google.com/patent/US20180213886A1/en
[41] H. Z. Adam, "Hockey stick," US1549971A, 1925. Accessed: May 2, 2025. [Online]. Available: https://patents.google.com/patent/US1549971A/en
[42] B. Dyer, "The controversy of sports technology: a systematic review," *Springerplus*, vol. 4, no. 1, pp. 1–12, Dec. 2015, doi: 10.1186/s40064-015-1331-x.
[43] F. Malizia and B. Blocken, "Bicycle aerodynamics: history, state-of-the-art and future perspectives," *Journal of Wind Engineering and Industrial Aerodynamics*, vol. 200, 2020, doi: 10.1016/j.jweia.2020.104134.
[44] F. Malizia and B. Blocken, "Cyclist aerodynamics through time: better, faster, stronger," *Journal of Wind Engineering and Industrial Aerodynamics*, vol. 214, 2021, doi: 10.1016/j.jweia.2021.104673.
[45] S. J. Haake, "The impact of technology on sporting performance in Olympic sports," *J Sports Sci*, vol. 27, no. 13, 2009, doi: 10.1080/02640410903062019.
[46] T. N. Crouch, D. Burton, Z. A. LaBry, and K. B. Blair, "Riding against the wind: a review of competition cycling aerodynamics," *Sports Engineering*, vol. 20, no. 2, 2017, doi: 10.1007/s12283-017-0234-1.
[47] C. Boardman, *Chris Boardman: The Biography of the Modern Bike: The Ultimate History of Bike Design*. London: Hachette UK, 2015.
[48] J. McPhee, "A review of dynamic models and measurements in golf," *Sports Engineering*, vol. 25, no. 1, 2022, doi: 10.1007/s12283-022-00387-0.
[49] M. R. Farrally *et al.*, "Golf science research at the beginning of the twenty-first century," *Journal of Sports Sciences*, 2003. doi: 10.1080/0264041031000102123.
[50] R. Stefani, "Olympic swimming gold: the suit or the swimmer in the suit?" *Significance*, vol. 9, no. 2, 2012, doi: 10.1111/j.1740-9713.2012.00553.x.
[51] S. R. Goodwill and S. J. Haake, "Ball spin generation for oblique impacts with a tennis racket," *Exp Mech*, vol. 44, no. 2, 2004, doi: 10.1177/0014485104039763.
[52] W. Fischer, "Tennis-racket," US4273331A, 1977. [Online]. Available: https://patents.google.com/patent/US4273331A/en
[53] G. T. Burns and D. P. Joubert, "Running shoes of the postmodern footwear era: a narrative overview of advanced footwear technology," *Int J Sports Physiol Perform*, vol. 19, no. 10, pp. 975–986, Aug. 2024, doi: 10.1123/IJSPP.2023-0446.
[54] "Do Nike's Zoom Vaporfly 4% marathon shoes actually make you run faster? | WIRED." Accessed: Mar. 21, 2025. [Online]. Available: https://www.wired.com/story/do-nike-zoom-vaporfly-make-you-run-faster/
[55] K. Hébert-Losier and M. Pamment, "Advancements in running shoe technology and their effects on running economy and performance – a current concepts overview," *Sports Biomech*, vol. 22, no. 3, pp. 335–350, 2023, doi: 10.1080/14763141.2022.2110512.
[56] B. Dyer, "Ten questions in sports engineering: supershoe use in distance running," *Sports Engineering*, vol. 27, no. 2, pp. 1–9, Dec. 2024, doi: 10.1007/s12283-024-00468-2.
[57] F. I. Michel *et al.*, "White Paper: functionality and efficacy of wrist protectors in snowboarding—towards a harmonized international standard," *Sports Engineering*, vol. 16, no. 4, 2013, doi: 10.1007/s12283-013-0113-3.
[58] T. J. Dickson and F. A. Terwiel, "Injury trends in alpine skiing and a snowboarding over the decade 2008–09 to 2017–18," *J Sci Med Sport*, vol. 24, no. 10, 2021, doi: 10.1016/j.jsams.2020.12.001.
[59] A. Natri, B. D. Beynnon, C. F. Ettlinger, R. J. Johnson, and J. E. Shealy, "Alpine ski bindings and injuries. Current findings," *Sports Medicine*, vol. 28, no. 1, pp. 35–48, Sep. 1999, doi: 10.2165/00007256-199928010-00004.
[60] "ISO 5355:2019 – Alpine ski-boots – Requirements and test methods." Accessed: May 8, 2025. [Online]. Available: https://www.iso.org/standard/74870.html
[61] "ISO 9462:2023 – Alpine ski-bindings – Requirements and test methods." Accessed: May 8, 2025. [Online]. Available: https://www.iso.org/standard/79640.html
[62] Richard Hymans, "Progression of world athletics records," 2020. Accessed: May 2, 2025. [Online]. Available: https://worldathletics.org/news/news/progression-of-world-athletics-records-on-sal

[63] B. Burkett, "Technology in Paralympic sport: performance enhancement or essential for performance?" *Br J Sports Med*, vol. 44, no. 3, 2010, doi: 10.1136/bjsm.2009.067249.

[64] S. Loland, "The ethics of performance-enhancing technology in sport," *J Philos Sport*, vol. 36, no. 2, p. 152, 2009, doi: 10.1080/00948705.2009.9714754.

[65] Y. Han, "Sports equipment: market data and analysis," Sep. 2024. Accessed: May 25, 2025. [Online]. Available: https://www.statista.com/study/48873/sports-equipment-market-data-and-analysis/

[66] S. Lüdemann, "Apparel Report 2019 – Sports & Swimwear Statista Consumer Market Outlook – Segment Report," 2019. [Online]. Available: https://www.statista.com/study/48858/apparel-report-sports-and-swimwear/

[67] M. Akin, "Athletic footwear: market data & analysis," 2023. [Online]. Available: https://www.statista.com/study/48863/athletic-footwear-report/

[68] Statistica, "Fitness trackers," 2024. [Online]. Available: https://www.statista.com/study/35598/fitness-and-activity-tracker/

[69] "Sports Engineering | EBSCO Research Starters." Accessed: Apr. 2, 2025. [Online]. Available: https://www.ebsco.com/research-starters/engineering/sports-engineering

[70] D. Ullman, *The Mechanical Design Process*, 5th ed. New York: McGraw-Hill, 2015.

[71] D. A. Gatenby and G. Foo, "Design for X (DFX): key to competitive, profitable products," *AT&T Technical Journal*, vol. 69, no. 3, pp. 2–13, May 1990, doi: 10.1002/j.1538-7305.1990.tb00332.x.

[72] I. M. Ilevbare, D. Probert, and R. Phaal, "A review of TRIZ, and its benefits and challenges in practice," *Technovation*, vol. 33, no. 2–3, pp. 30–37, Feb. 2013, doi: 10.1016/j.technovation.2012.11.003.

[73] D. G. Ullman, *The Mechanical Design Process*, 6th ed. New York: McGraw-Hill Education, 2017.

[74] K. T. Ulrich, S. D. Eppinger, and M. C. Yang, *Product Design and Development*, 7th ed. New York: McGraw-Hill Education, 2019.

[75] "Complete List – The 50 Best Inventions of 2009 – TIME." Accessed: Jul. 9, 2025. [Online]. Available: https://content.time.com/time/specials/packages/completelist/0,29569,1934027,00.html

[76] C. Mortimer and J. King, "Yike Bike: Eco-innovation Business Case Study 1," Auckland, 2011. Accessed: May 30, 2025. [Online]. Available: https://www.landcareresearch.co.nz/assets/researchpubs/case_study_yike_bike.pdf

[77] S. J. Mackenzie, D. Ura, E. M. Henrikson, and L. Kell, "Influence of putter offset on performance," *Sports Engineering*, vol. 25, no. 1, p. 15, 2022, doi: 10.1007/s12283-022-00379-0.

[78] A. S. McIntosh, "Biomechanical considerations in the design of equipment to prevent sports injury," *Proc Inst Mech Eng P J Sport Eng Technol*, vol. 226, no. 3–4, pp. 193–199, Dec. 2011, doi: 10.1177/1754337111431024.

[79] N. Wilson, A. Thomson, and P. Riches, "Development and presentation of the first design process model for sports equipment design," *Res Eng Des*, vol. 28, no. 4, pp. 495–509, 2017, doi: 10.1007/s00163-017-0257-4.

[80] "US959053A – Golf-club. – Google Patents." Accessed: Aug. 19, 2025. [Online]. Available: https://patents.google.com/patent/US959053A/en

[81] M. A. Runco, "Cognition and creativity," *Creativity*, pp. 1–36, Jan. 2023, doi: 10.1016/B978-0-08-102617-5.00012-6.

[82] R. Sham, A. Abdhul, and S. Reddy, "Innovative product design with generative AI," in *Minds Unveiled*, R. V. Rodriguez and K. Hemachandran, Eds, New York: Productivity Press, 2024, pp. 137–157.

[83] S. Robertson *et al.*, "Development of a sports technology quality framework," *J Sports Sci*, vol. 41, no. 22, pp. 1983–1993, Nov. 2023, doi: 10.1080/02640414.2024.2308435.

[84] Sports Technology Research Network (STRN), "A New Sports Tech Quality Framework White Paper: Bringing the Framework to Life – Sports Tech Research Network." Accessed: Apr. 4, 2025. [Online]. Available: https://strn.co/news-insights/2025/01/13/a-new-sports-tech-quality-framework-white-paper-bringing-the-framework-to-life

[85] E. Mencarini, C. Leonardi, A. Cappelletti, D. Giovanelli, A. De Angeli, and M. Zancanaro, "Co-designing wearable devices for sports: the case study of sport climbing," *International Journal of Human Computer Studies*, vol. 124, pp. 26–43, Apr. 2019, doi: 10.1016/j.ijhcs.2018.10.005.

[86] F. K. Fuss, A. Subic, M. Strangwood, and R. Mehta, *Routledge Handbook of Sports Technology and Engineering*, 1st ed. London: Routledge, 2013.
[87] C. Matthews, *Engineers' Data Book*, 4th ed. Hoboken, NJ: Wiley-Blackwell, 2012.
[88] K. P. Clark, L. J. Ryan, and P. G. Weyand, "A general relationship links gait mechanics and running ground reaction forces," *Journal of Experimental Biology*, Jan. 2016, doi: 10.1242/jeb.138057.
[89] B. T. van Oeveren, C. J. de Ruiter, P. J. Beek, and J. H. van Dieën, "The biomechanics of running and running styles: a synthesis," *Sports Biomech*, vol. 23, no. 4, pp. 516–554, Apr. 2024, doi: 10.1080/14763141.2021.1873411.
[90] H. M. Rice, S. T. Jamison, and I. S. Davis, "Footwear matters," *Med Sci Sports Exerc*, vol. 48, no. 12, pp. 2462–2468, Dec. 2016, doi: 10.1249/MSS.0000000000001030.
[91] J. Turner, S. E. Forrester, A. C. Mears, and J. R. Roberts, "Reliability of repeat golf club testing sessions with modified club moment of inertia," *Sports Engineering*, vol. 23, no. 1, p. 8, Dec. 2020, doi: 10.1007/s12283-020-0320-7.
[92] R. Cross, *Physics of Baseball and Softball*, 1st ed. New York: Springer, 2011.
[93] D. Whiteside, B. Elliott, B. Lay, and M. Reid, "The effect of racquet swing weight on serve kinematics in elite adolescent female tennis players," *J Sci Med Sport*, vol. 17, no. 1, 2014, doi: 10.1016/j.jsams.2013.03.001.
[94] R. Cross and R. Bower, "Effects of swing-weight on swing speed and racket power," *J Sports Sci*, vol. 24, no. 1, 2006, doi: 10.1080/02640410500127876.
[95] D. Schorah, S. Choppin, and D. James, "Effects of moment of inertia on restricted motion swing speed," *Sports Biomech*, vol. 14, no. 2, 2015, doi: 10.1080/14763141.2015.1027949.
[96] H. Brody, "The moment of inertia of a tennis racket," *Phys Teach*, vol. 23, no. 4, Apr. 1985, doi: 10.1119/1.2341781.
[97] R. Cross and A. M. Nathan, "Performance versus moment of inertia of sporting implements," *Sports Technology*, vol. 2, no. 1–2, 2009, doi: 10.1002/jst.88.
[98] S. R. Mitchell, R. Jones, and M. King, "Head speed vs. racket inertia in the tennis serve," *Sports Engineering*, vol. 3, no. 2, 2000, doi: 10.1046/j.1460-2687.2000.00051.x.
[99] J. Frère, M. L'Hermette, J. Slawinski, and C. Tourny-Chollet, "Mechanics of pole vaulting: a review," *Sports Biomech*, vol. 9, no. 2, 2010, doi: 10.1080/14763141.2010.492430.
[100] A. Haron and K. A. Ismail, "Coefficient of restitution of sports balls: a normal drop test," *IOP Conf Ser Mater Sci Eng*, vol. 36, pp. 1–8, 2012, doi: 10.1088/1757-899X/36/1/012038.
[101] International Tennis Federation (ITF), "ITF Rules of Tennis," 2025. Accessed: May 2, 2025. [Online]. Available: https://www.itftennis.com/en/about-us/governance/rules-and-regulations/
[102] Fédération Internationale de Football Association (FIFA), "FIFA Quality Programme for Footballs (outdoor, futsal and beach soccer footballs) Test Manual," Zurich, Dec. 2023. Accessed: May 4, 2025. [Online]. Available: https://inside.fifa.com/innovation/standards/footballs
[103] R. Cross, "Impact of sports balls with striking implements," *Sports Engineering*, vol. 17, no. 1, pp. 3–22, Mar. 2014, doi: 10.1007/S12283-013-0132-0.
[104] R. J. Elkington, J. L. Armitage, T. Khan, and M. G. Bryant, "Sticky feet: a tribological study of climbing shoe rubber," *Sports Engineering*, vol. 27, no. 2, p. 30, Dec. 2024, doi: 10.1007/s12283-024-00474-4.
[105] J. W. Wannop, J. T. Worobets, and D. J. Stefanyshyn, "Footwear traction and lower extremity joint loading," *Am J Sports Med*, vol. 38, no. 6, pp. 1221–1228, Jun. 2010, doi: 10.1177/0363546509359065.
[106] J. W. Wannop, G. Luo, and D. J. Stefanyshyn, "Footwear traction at different areas on artificial and natural grass fields," *Sports Engineering*, vol. 15, no. 2, pp. 111–116, Jun. 2012, doi: 10.1007/s12283-012-0091-x.
[107] T. Allen, S. Haake, and S. Goodwill, "Effect of friction on tennis ball impacts," *Proc Inst Mech Eng P J Sport Eng Technol*, vol. 224, no. 3, 2010, doi: 10.1243/17543371JSET66.
[108] E. Goff, *Gold Medal Physics: The Science of Sports*, 1st ed., vol. 1. Baltimore, MD: Johns Hopkins University Press, 2010.
[109] M. Denny, *Gliding for Gold: The Physics of Winter Sports*. Baltimore, MD: Johns Hopkins University Press, 2011.
[110] A. T. Morales, J. A. Tamayo Fajardo, and H. González-García, "High-speed swimsuits and their historical development in competitive swimming," *Front Psychol*, vol. 10, p. 480719, Dec. 2019, doi: 10.3389/fpsyg.2019.02639.

[111] L. M. O'Connor and J. A. Vozenilek, "Is it the athlete or the equipment? An analysis of the top swim performances from 1990 to 2010," *J Strength Cond Res*, vol. 25, no. 12, pp. 3239–3241, Dec. 2011, doi: 10.1519/JSC.0B013E3182392C5F.

[112] L. Brownlie, "Aerodynamic drag reduction in winter sports: the quest for 'free speed,'" *Proc Inst Mech Eng P J Sport Eng Technol*, vol. 235, no. 4, pp. 365–404, Dec. 2021, doi: 10.1177/1754337120921091.

[113] O. Elfmark, K. E. Teigen Giljarhus, F. F. Liland, L. Oggiano, and R. Reid, "Aerodynamic investigation of tucked positions in alpine skiing," *J Biomech*, vol. 119, p. 110327, Apr. 2021, doi: 10.1016/J.JBIOMECH.2021.110327.

[114] A. Winkler and A. Pernpeintner, "Improving the performance of a bobsleigh by aerodynamic optimization," in *The Engineering of Sport 7*, M. Estivalet and P. Brisson, Eds, Paris: Springer Paris, 2009, pp. 329–338. doi: 10.1007/978-2-287-99056-4_41.

[115] A. J. Ward-Smith, "Air resistance and its influence on the biomechanics and energetics of sprinting at sea level and at altitude," *J Biomech*, vol. 17, no. 5, pp. 339–347, Jan. 1984, doi: 10.1016/0021-9290(84)90028-9.

[116] F. White, *Fluid Mechanics*, 7th ed. New York: McGraw-Hill, 2011.

[117] B. S. Massey and J. Ward-Smith, *Mechanics of Fluids*, 7th ed. London: CRC Press, 1998.

[118] T. F. Irvine and J. P. Hartnett, *Steam and Air Tables in SI Units: Including Data for Other Substances with a Separate Mollier Chart for Steam*. London: Hemisphere Publishing, 1976, p. 127.

[119] D. Bernoulli, *Hydrodynamica*. Strasbourg: Johann Reinhold Dulsecker, 1738.

[120] S. F. Hoerner, *Fluid-Dynamic Drag*, Hoerner Fluid Dynamics, Brick Town, New Jersey. 1965.

[121] E. Achenbach, "Influence of surface roughness on the cross-flow around a circular cylinder," *J Fluid Mech*, vol. 46, no. 2, pp. 321–335, Mar. 1971, doi: 10.1017/S0022112071000569.

[122] E. Achenbach, "Experiments on the flow past spheres at very high Reynolds numbers," *J Fluid Mech*, vol. 54, no. 3, pp. 565–575, Aug. 1972, doi: 10.1017/S0022112072000874.

[123] J. E. Goff, C. M. Hobson, T. Asai, and S. Hong, "Wind-tunnel experiments and trajectory analyses for five nonspinning soccer balls," in *The Engineering of Sport 11*, A. Jansen, Ed., Delft: Procedia Engineering by Elsevier, Jan. 2016, pp. 32–37. doi: 10.1016/J.PROENG.2016.06.185.

[124] H. Higuchi and T. Kiura, "Aerodynamics of knuckle ball: flow-structure interaction problem on a pitched baseball without spin," *J Fluids Struct*, vol. 32, pp. 65–77, Jul. 2012, doi: 10.1016/J.JFLUIDSTRUCTS.2012.01.004.

[125] J. E. Goff, "A review of recent research into aerodynamics of sport projectiles," *Sports Engineering*, vol. 16, no. 3, pp. 137–154, Apr. 2013, doi: 10.1007/S12283-013-0117-Z.

[126] B. Lyu, J. Kensrud, and L. Smith, "The reverse Magnus effect in golf balls," *Sports Engineering*, vol. 23, no. 1, p. 3, Dec. 2020, doi: 10.1007/s12283-020-0318-1.

[127] N. Sakib and B. L. Smith, "Study of the reverse Magnus effect on a golf ball and a smooth ball moving through still air," *Exp Fluids*, vol. 61, no. 5, p. 115, May 2020, doi: 10.1007/s00348-020-02946-2.

[128] J. A. Scobie, W. P. Shelley, R. W. Jackson, S. P. Hughes, and G. D. Lock, "Practical perspective of cricket ball swing," *Proc Inst Mech Eng P J Sport Eng Technol*, vol. 234, no. 1, pp. 59–71, Mar. 2020, doi: 10.1177/1754337119872874.

[129] J. Anderson, *Fundamentals of Aerodynamics*, 6th ed. New York: McGraw Hill, 2016.

[130] M. Ashby, *Materials Selection in Mechanical Design*, 4th ed. Boston, MA: Butterworth-Heinemann, 2010.

[131] M. Strangwood, A. D. G. Johnson, and S. R. Otto, "Energy losses in viscoelastic golf balls," *Proceedings of the Institution of Mechanical Engineers, Part L: Journal of Materials: Design and Applications*, vol. 220, no. 1, pp. 23–30, Jan. 2006, doi: 10.1243/146442005X34548.

[132] International Bobsleigh & Skeleton Federation (IBSF), "International Bobsleigh Rules 2024," Sep. 2024. Accessed: May 4, 2025. [Online]. Available: https://www.ibsf.org/en/inside-ibsf/downloads-resources

[133] M. Jenkins, *Materials in Sports Equipment*, 1st ed. Cambridge: Woodhead Publishing, 2003.

[134] A. Subic, *Materials in Sports Equipment*, 2nd ed. Cambridge: Woodhead Publishing, 2007.

[135] Society of Automobile Engineers (SAE), *Composite Materials Handbook – Polymer Matrix Composites: Material Properties*, vol. 2, June. SAE International, 2018.

[136] G. H. Staab, *Laminar Composites*, 2nd ed. Boston, MA: Butterworth-Heinemann, 2015.

[137] "ISO 527-4:2023(en), Plastics – Determination of tensile properties – Part 4: Test conditions for isotropic and orthotropic fibre-reinforced plastic composites." Accessed: Jul. 24, 2025. [Online]. Available: https://www.iso.org/obp/ui/fr/#iso:std:iso:527:-4:ed-3:v1:en

[138] E. M. C. Jones *et al.*, "A good practices guide for digital image correlation," International Digital Image Correlation Society, Aug. 2018. doi: 10.32720/idics/gpg.ed1/print.format.

[139] L. Mullins, "Softening of rubber by deformation," *Rubber Chemistry and Technology*, vol. 42, no. 1, 1969, doi: 10.5254/1.3539210.

[140] N. Bailly, Y. Petit, J. M. Desrosier, O. Laperriere, S. Langlois, and E. Wagnac, "Strain rate dependent behavior of vinyl nitrile helmet foam in compression and combined compression and shear," *Applied Sciences*, vol. 10, no. 22, p. 8286, Nov. 2020, doi: 10.3390/APP10228286.

[141] S. D. Burbank and L. V. Smith, "Dynamic characterization of rigid foam used in finite element sports ball simulations," *Proc Inst Mech Eng P J Sport Eng Technol*, vol. 226, no. 2, pp. 77–85, Jun. 2012, doi: 10.1177/1754337112441112.

[142] A. E. F. Athanasiadis, M. A. Dias, and M. K. Budzik, "Can confined mechanical metamaterials replace adhesives?" *Extreme Mech Lett*, vol. 48, p. 101411, Oct. 2021, doi: 10.1016/j.eml.2021.101411.

[143] S. R. Goodwill, R. Kirk, and S. J. Haake, "Experimental and finite element analysis of a tennis ball impact on a rigid surface," *Sports Engineering*, vol. 8, no. 3, pp. 145–158, Jan. 2005, doi: 10.1007/BF02844015.

[144] D. S. Price, R. Jones, and A. R. Harland, "Advanced finite-element modelling of a 32-panel soccer ball," *Proc Inst Mech Eng C J Mech Eng Sci*, vol. 221, no. 11, 2007, doi: 10.1243/09544062JMES711.

[145] S. A. Imam *et al.*, "Finite element model to simulate impact on a soft tissue simulant," *Sports Engineering*, vol. 26, no. 1, 2023, doi: 10.1007/s12283-023-00407-7.

[146] Y. Song, E. Shao, I. Bíró, J. S. Baker, and Y. Gu, "Finite element modelling for footwear design and evaluation: a systematic scoping review," *Heliyon*, vol. 8, no. 10, p. e10940, Oct. 2022, doi: 10.1016/j.heliyon.2022.e10940.

[147] T. Shepherd, K. Winwood, P. Venkatraman, A. Alderson, and T. Allen, "Validation of a finite element modeling process for auxetic structures under impact," *Phys Status Solidi B Basic Res*, vol. 257, no. 10, 2020, doi: 10.1002/pssb.201900197.

[148] K. Tanaka and K. Sekizawa, "Construction of a finite element model of golf clubs and influence of shaft stiffness on its dynamic behavior," in *The 12th Conference of the International Sports Engineering Association*, Basel Switzerland: MDPI, Feb. 2018, p. 247. doi: 10.3390/proceedings2060247.

[149] T. Allen, S. Haake, and S. Goodwill, "Comparison of a finite element model of a tennis racket to experimental data," *Sports Engineering*, vol. 12, no. 2, pp. 87–98, 2009, doi: 10.1007/s12283-009-0032-5.

[150] R. C. Hibbeler, *Mechanics of Materials*, 10th ed., vol. 41, no. 1066. Harlow: Pearson, 2018.

[151] J. E. Gordon, *The New Science of Strong Materials or Why You Don't Fall Through the Floor*. London: Penguin, 1991.

[152] M. Dorsemaine, I. S. Scher, T. Allen, C. Masson, L. L. Stepan, and P.-J. Arnoux, "Recommendations to improve ski area safety with obstacle padding," *JSAMS Plus*, vol. 2, 2023, doi: 10.1016/j.jsampl.2023.100036.

[153] G. T. Burns and D. P. Joubert, "Running shoes of the postmodern footwear era: a narrative overview of advanced footwear technology," *Int J Sports Physiol Perform*, pp. 1–12, 2024, doi: 10.1123/ijspp.2023-0446.

[154] W. Hoogkamer, S. Kipp, J. H. Frank, E. M. Farina, G. Luo, and R. Kram, "A comparison of the energetic cost of running in marathon racing shoes," *Sports Medicine*, vol. 48, no. 4, 2018, doi: 10.1007/s40279-017-0811-2.

[155] L. J. Gibson and M. F. Ashby, *Cellular Solids: Structure and Properties*, 2nd edn. Cambridge: Cambridge University Press, 2014. doi: 10.1017/CBO9781139878326.

[156] Q. Li *et al.*, "Contemporary advances in polymer applications for sporting goods: fundamentals, properties, and applications," *RSC Adv*, vol. 14, no. 50, pp. 37445–37469, Nov. 2024, doi: 10.1039/D4RA06544A.

[157] M. Parisi, T. Allen, M. Colonna, N. Pugno, and O. Duncan, "Indentation and impact response of conventional, auxetic, and shear thickening gel infused auxetic closed cell foam," *Smart Mater Struct*, vol. 32, no. 7, 2023, doi: 10.1088/1361-665X/acd91c.

[158] M. Parisi, G. La Fauci, N. M. Pugno, and M. Colonna, "Use of shear thickening fluids in sport protection applications: a review," *Frontiers in Materials*, vol. 10, 2023, doi: 10.3389/fmats.2023.1285995.

[159] M. Zou *et al.*, "Elastic shear-stiffening composites with locally tunable mechanics for protection and damping," *Appl Mater Today*, vol. 40, p. 102396, Oct. 2024, doi: 10.1016/j.apmt.2024.102396.

[160] G. La Fauci, M. Parisi, A. Nanni, L. Crosetta, N. M. Pugno, and M. Colonna, "Design and proof-of-concept of an advanced protective system for the dissipation of tangential impact energy in helmets, based on non-Newtonian fluids," *Smart Mater Struct*, vol. 32, no. 4, 2023, doi: 10.1088/1361-665X/acc148.

[161] O. Duncan, "Fast optimisation of honeycombs for impact protection," *Int J Impact Eng*, vol. 189, p. 104973, Jul. 2024, doi: 10.1016/j.ijimpeng.2024.104973.

[162] V. L. Popov, M. Heß, and E. Willert, *Handbook of Contact Mechanics: Exact Solutions of Axisymmetric Contact Problems*. Berlin: Springer, 2019.

[163] K. L. Alderson, V. R. Simkins, V. L. Coenen, P. J. Davies, A. Alderson, and K. E. Evans, "How to make auxetic fibre reinforced composites," *Phys Status Solidi B Basic Res*, vol. 242, no. 3, 2005, doi: 10.1002/pssb.200460371.

[164] N. Liu, N. Mehreganian, and P. Sareh, "Never better than 5/6: the fundamental limit of energy absorption efficiency for negative-stiffness curved-beam honeycombs," *Mater Des*, vol. 243, p. 113024, Jul. 2024, doi: 10.1016/j.matdes.2024.113024.

[165] J. Dunne, R. D. Crapnell, K. K. Dudek, T. Allen, C. E. Banks, and O. Duncan, "Electro-thermally controlled active mechanical metamaterials with programmable stiffness and non-reciprocity," *Advanced Science*, p. e11669, Aug. 2025, doi: 10.1002/ADVS.202511669.

[166] S. Janbaz, K. Narooei, T. Van Manen, and A. A. Zadpoor, "Strain rate-dependent mechanical metamaterials," *Sci Adv*, vol. 6, no. 25, 2020, doi: 10.1126/sciadv.aba0616.

[167] S. G. Hayes and P. Venkatraman, *Materials and Technology for Sportswear and Performance Apparel*, 1st ed. Boca Raton, FL: CRC Press, 2015.

[168] A. Subic, B. Shabani, M. Hedayati, and E. Crossin, "Capability framework for sustainable manufacturing of sports apparel and footwear," *Sustainability*, vol. 4, no. 9, pp. 2127–2145, 2012, doi: 10.3390/su4092127.

[169] R. Thompson, *Manufacturing Processes for Design Professionals*. London: Thames & Hudson, 2007.

[170] S. Kalpakjian and S. Schmid, *Manufacturing, Engineering and Technology*, 8th ed. London: Pearson, 2022.

[171] "The Bat Law | MCC." Accessed: Jun. 6, 2025. [Online]. Available: https://www.lords.org/mcc/the-laws-of-cricket/the-bat

[172] "Advanced manufacturing – English willow cricket bats." Accessed: May 2, 2025. [Online]. Available: https://b3cricket.com/pages/advanced-manufacturing

[173] D. Curtis, B. Heller, and T. Senior, "Methods for estimating moment of inertia of cricket bats," *Sports Engineering*, vol. 24, no. 1, pp. 1–8, Dec. 2021, doi: 10.1007/s12283-021-00349-y.

[174] T. Allen, O. Fauteux-Brault, D. James, and D. Curtis, "Finite element model of a cricket ball impacting a bat," *Procedia Engineering*, vol. 72, pp. 521–526, 2014. doi: 10.1016/j.proeng.2014.06.090.

[175] J. I. Novak and A. R. Novak, "Is additive manufacturing improving performance in Sports? A systematic review," *Proc Inst Mech Eng P J Sport Eng Technol*, vol. 235, no. 3, pp. 163–175, 2021, doi: 10.1177/1754337120971521.

[176] E. Pei *et al.*, Eds, *Springer Handbook of Additive Manufacturing*. Cham: Springer, 2023.

[177] Y. Song, Y. Ghafari, A. Asefnejad, and D. Toghraie, "An overview of selective laser sintering 3D printing technology for biomedical and sports device applications: processes, materials, and applications," *Opt Laser Technol*, vol. 171, p. 110459, 2024, doi: https://doi.org/10.1016/j.optlastec.2023.110459.

[178] A. Saboori, A. Aversa, G. Marchese, S. Biamino, M. Lombardi, and P. Fino, "Application of directed energy deposition-based additive manufacturing in repair," *Applied Sciences*, vol. 9, no. 16, p. 3316, Aug. 2019, doi: 10.3390/APP9163316.

[179] S. J. Raykar, M. M. Narke, S. B. Desai, and S. S. Warke, "Manufacturing of 3D printed sports helmet," in *Techno-Societal 2018 – Proceedings of the 2nd International Conference on Advanced Technologies for Societal Applications*, P. M. Pawar, B. P. Ronge, R. Balasubramaniam, A. S. Vibhute, and S. S. Apte, Eds, Cham: Springer, 2020, pp. 771–778. doi: 10.1007/978-3-030-16962-6_77.

[180] J. Van Tubergen, J. Verlinden, M. Stroober, and R. Baldewsing, "Suited for performance: fast full-scale replication of athlete with FDM," in *Proceedings – SCF 2017: ACM Symposium on Computational Fabrication*, S. Mueller, S. Coros, and N. Fang, Eds, New York: Association for Computing Machinery, Inc, Jun. 2017. doi: 10.1145/3083157.3092888.

[181] N. G. Tanikella, B. Wittbrodt, and J. M. Pearce, "Tensile strength of commercial polymer materials for fused filament fabrication 3D printing," *Addit Manuf*, vol. 15, pp. 40–47, May 2017, doi: 10.1016/J.ADDMA.2017.03.005.

[182] S. A. M. Tofail, E. P. Koumoulos, A. Bandyopadhyay, S. Bose, L. O'Donoghue, and C. Charitidis, "Additive manufacturing: scientific and technological challenges, market uptake and opportunities," *Materials Today*, vol. 21, no. 1, pp. 22–37, Jan. 2018, doi: 10.1016/J.MATTOD.2017.07.001.

[183] "AusCycling statement – Tokyo 2020 Olympic Games handlebar investigation findings | AusCycling." Accessed: Apr. 29, 2025. [Online]. Available: https://auscycling.org.au/news/auscycling-statement-tokyo-2020-olympic-games-handlebar-investigation-findings

[184] O. Diegel, A. Nordin, and D. Motte, "DfAM strategic design considerations," in *Springer Series in Advanced Manufacturing*, Berlin: Springer Nature, 2019, pp. 41–70. doi: 10.1007/978-981-13-8281-9_3.

[185] M. Andreff and W. Andreff, "Global trade in sports goods: international specialisation of major trading countries," *European Sport Management Quarterly*, vol. 9, no. 3, pp. 259–294, Sep. 2009, doi: 10.1080/16184740903024029.

[186] M. E. Dunn, M. Mills, and D. Veríssimo, "Evaluating the impact of the documentary series Blue Planet on viewers' plastic consumption behaviors," *Conserv Sci Pract*, vol. 2, no. 10, p. e280, 2020, doi: https://doi.org/10.1111/csp2.280.

[187] A. Subie, A. Mouritz, and O. Troynikov, "Sustainable design and environmental impact of materials in sports products," *Leisure/Loisir*, vol. 2, no. 3–4, pp. 67–79, Jan. 2009, doi: 10.1080/19346182.2009.9648504.

[188] "EU Ecolabel – Home." Accessed: Sep. 2, 2025. [Online]. Available: https://environment.ec.europa.eu/topics/circular-economy/eu-ecolabel_en

[189] H. McGowan, P. Fleming, D. James, J. McMahon, J. H. Pak, and S. Forrester, "Investigating normal stress effects on the shear and traction characteristics of performance infill materials used in artificial turf surfaces," *Sports Engineering*, vol. 28, no. 1, pp. 1–14, Jun. 2025, doi: 10.1007/S12283-025-00489-5.

[190] D. Cole *et al.*, "Comparison of player perceptions to mechanical measurements of third generation synthetic turf football surfaces," *Sports Engineering*, vol. 26, no. 1, p. 5, Dec. 2023, doi: 10.1007/s12283-022-00398-x.

[191] D. Cole *et al.*, "Development and evaluation of a sensory panel for collecting reliable player perceptions of third-generation synthetic turf football surfaces," *Sports Engineering*, vol. 26, no. 1, 2023, doi: 10.1007/s12283-022-00395-0.

[192] S. Joshi and S. Temgire, "Life below water and the fate of humanity: with special reference to the efforts by adidas towards reducing the effects of plastic waste on the marine life," *Sustainability, Agri, Food and Environmental Research-DISCONTINUED*, vol. 10, no. X, p. 2022, Jan. 2022, doi: 10.7770/SAFER-V10N1-ART2559.

[193] V. E. Annamalai, S. Vijayan, H. Saeedipour, and K. L. Goh, "The rise of short fibre reinforced plastics," *Reinforced Plastics*, vol. 64, no. 2, pp. 97–102, 2020, doi: https://doi.org/10.1016/j.repl.2019.07.002.

[194] T. P. Lyon and J. W. Maxwell, "Greenwash: corporate environmental disclosure under threat of audit," *J Econ Manag Strategy*, vol. 20, no. 1, pp. 3–41, Mar. 2011, doi: https://doi.org/10.1111/j.1530-9134.2010.00282.x.

[195] N. Sharma and A. Paço, "Moral disengagement: a guilt free mechanism for non-green buying behavior," *J Clean Prod*, vol. 297, p. 126649, 2021, doi: https://doi.org/10.1016/j.jclepro.2021.126649.

[196] C. Podara *et al.*, "Recent trends of recycling and upcycling of polymers and composites: a comprehensive review," *Recycling*, vol. 9, no. 3, p. 37, 2024, doi: 10.3390/recycling9030037.

[197] Renishaw, "First metal 3D printed bicycle frame manufactured by Renishaw for Empire Cycles." Accessed: Jun. 6, 2025. [Online]. Available: https://www.renishaw.com/en/first-metal-3d-printed-bicycle-frame-manufactured-by-renishaw-for-empire-cycles-24154?srsltid=AfmBOoq61lqdASJvtsB8QYZgkuC46_6P0hSM4IjUdyRmEz1V0viXSUCu
[198] O. Diegel, A. Nordin, and D. Motte, *A Practical Guide to Design for Additive Manufacturing*. Singapore: Springer, 2019.
[199] W. Hall and Z. Javanbakht, *Design and Manufacture of Fibre-Reinforced Composites*, vol. 158. Cham: Springer, 2021.
[200] E. Tempelman, H. Shercliff, and B. Ninaber van Eyben, *Manufacturing and Design*. Oxford: Elsevier, 2014.
[201] "ISO 4210-6:2023 – Cycles – Safety requirements for bicycles – Part 6: Frame and fork test methods." Accessed: Apr. 30, 2025. [Online]. Available: https://www.iso.org/standard/78081.html
[202] "ISO 20320:2020 – Protective clothing for use in snowboarding – Wrist protectors – Requirements and test methods." Accessed: Apr. 30, 2025. [Online]. Available: https://www.iso.org/standard/67665.html
[203] World Rugby, "Regulations | World Rugby." Accessed: May 25, 2025. [Online]. Available: https://www.world.rugby/organisation/governance/regulations/reg-12
[204] International Ski and Snowboard Federation (FIS), "Specifications for alpine competition equipment," Oberhofen, Jun. 2024. [Online]. Available: www.fis-ski.com
[205] "ISO 4980:2023 – Benefit-risk assessment for sports and recreational facilities, activities and equipment." Accessed: May 1, 2025. [Online]. Available: https://www.iso.org/standard/80573.html
[206] EN, "EN 1078:2012 – Helmets for pedal cyclists and for users of skateboards and roller skates." Accessed: Jul. 21, 2025. [Online]. Available: https://standards.iteh.ai/catalog/standards/cen/c9cc1a5e-cfa9-42b4-82a9-700d4bcd2dad/en-1078-2012?srsltid=AfmBOoow_7nL8JRFf-6hT8UBVS41umtJ-QHOyr-yATaocN3pPfocWsdb
[207] S. Odenwald, "Test methods in the development of sports equipment," in *The Engineering of Sport 6*, E. F. Moritz and S. Haake, Eds, New York: Springer, 2006, pp. 301–306.
[208] T. B. Allen, S. J. Haake, and S. R. Goodwill, "Effect of tennis racket parameters on a simulated groundstroke," *J Sports Sci*, vol. 29, no. 3, 2011, doi: 10.1080/02640414.2010.526131.
[209] J. R. Lee, A. Harland, J. Roberts, D. Price, and G. Jadoul, "The interdependence of functional properties of a football boot outsole during the shape optimisation process," *Sports Engineering*, vol. 27, no. 2, p. 29, Dec. 2024, doi: 10.1007/s12283-024-00464-6.
[210] B. Blocken *et al.*, "Aerodynamic drag in cycling pelotons: new insights by CFD simulation and wind tunnel testing," *Journal of Wind Engineering and Industrial Aerodynamics*, vol. 179, pp. 319–337, Aug. 2018, doi: 10.1016/J.JWEIA.2018.06.011.
[211] V. Senner and M. Caine, "Physical (mechanical) models for sports equipment research, development and testing," *Proc Inst Mech Eng P J Sport Eng Technol*, vol. 237, no. 1, pp. 3–6, Mar. 2023, doi: 10.1177/17543371231158896.
[212] B. Emsley, J. Farmer, P. Sherratt, P. Goodall, T. Jackson, and A. West, "An overview of the test methodology used in current cycling helmet standards and literature," *Int J Impact Eng*, vol. 188, p. 104928, Jun. 2024, doi: 10.1016/j.ijimpeng.2024.104928.
[213] S. Schwanitz and S. Odenwald, "Long-term cushioning properties of running shoes," in *The Engineering of Sport 7*, M. Estivalet and P. Brisson, Eds, Paris: Springer, Jun. 2008.
[214] ASTM International, "F1614 standard test method for shock attenuating properties of materials systems for athletic footwear." Accessed: May 25, 2025. [Online]. Available: https://store.astm.org/f1614-99.html
[215] B. Lane, P. Sherratt, X. Hu, and A. Harland, "Measurement of strain and strain rate during the impact of tennis ball cores," *Applied Sciences (Switzerland)*, vol. 8, no. 3, 2018, doi: 10.3390/app8030371.
[216] D. Knudson, *Fundamentals of Biomechanics*, 3rd ed. Cham: Springer, 2021. doi: 10.1007/978-3-030-51838-7.
[217] R. Bartlett, *Introduction to Sports Biomechanics: Analysing Human Movement Patterns*, 3rd ed. London: Routledge, 2014. doi: 10.4324/9781315889504.
[218] R. Bartlett and M. Bussey, *Sports Biomechanics: Reducing Injury Risk and Improving Sports Performance*, 2nd ed. London: Routledge, 2013.

[219] S. J. Dixon, "Use of pressure insoles to compare in-shoe loading for modern running shoes," *Ergonomics*, vol. 51, no. 10, pp. 1503–1514, Oct. 2008, doi: 10.1080/00140130802239562.
[220] D. C. Low and S. J. Dixon, "Footscan pressure insoles: accuracy and reliability of force and pressure measurements in running," *Gait Posture*, vol. 32, no. 4, pp. 664–666, Oct. 2010, doi: 10.1016/j.gaitpost.2010.08.002.
[221] A. C. Mears, E. M. P. Williams, P. W. Kong, P. Wood, K. Brubacher, and T. Allen, "Towards inclusive sports engineering research: considerations of research methods," *Sports Engineering*, vol. 27, no. 2, p. 32, Dec. 2024, doi: 10.1007/s12283-024-00472-6.
[222] E. M. Winter and R. J. Maughan, "Requirements for ethics approvals," *J Sports Sci*, vol. 27, no. 10, p. 985, 2009, doi: 10.1080/02640410903178344.
[223] C. Adams, D. James, T. Senior, T. Allen, and N. Hamilton, "Effect of surrogate design on the measured stiffness of snowboarding wrist protectors," *Sports Engineering*, vol. 21, no. 3, 2018, doi: 10.1007/s12283-018-0266-1.
[224] G. E. Leslie, K. Winwood, W. Wang, N. Hamilton, and T. Allen, "Effect of limb surrogate surface compliance on the impact response of wrist protectors," *JSAMS Plus*, vol. 2, p. 100023, 2023, doi: 10.1016/j.jsampl.2023.100023.
[225] G. Leslie, W. Wang, K. Winwood, C. Liauw, N. Hamilton, and T. Allen, "Effect of surrogate surface compliance on the measured stiffness of snowboarding wrist protectors," *Proceedings*, vol. 49, no. 1, p. 84, 2020. doi: 10.3390/proceedings2020049084.
[226] N. Petrone *et al.*, "Feasibility of using a novel instrumented human head surrogate to measure helmet, head and brain kinematics and intracranial pressure during multidirectional impact tests," *J Sci Med Sport*, vol. 22, pp. S78–S84, Aug. 2019, doi: 10.1016/j.jsams.2019.05.015.
[227] S. A. Imam *et al.*, "A finite element model for predicting impact-induced damage to a skin simulant," *Sci Rep*, vol. 14, no. 1, p. 12810, Jun. 2024, doi: 10.1038/s41598-024-60369-z.
[228] T. Payne, S. Mitchell, B. Halkon, R. Bibb, and M. Waters, "Development of a synthetic human thigh impact surrogate for sports personal protective equipment testing," *Proc Inst Mech Eng P J Sport Eng Technol*, vol. 230, no. 1, pp. 5–16, Mar. 2016, doi: 10.1177/1754337115582294.
[229] S. Mitchell, T. Payne, and R. Bibb, "Design of human surrogates for the study of biomechanical injury: a review," *Crit Rev Biomed Eng.*, vol. 41, no. 1, pp. 51–89, 2013, doi: 10.1615/CritRevBiomedEng.2013006847.
[230] C. Adams, T. Allen, T. Senior, D. James, and N. Hamilton, "Impact testing of snowboarding wrist protectors," *Proc Inst Mech Eng P J Sport Eng Technol*, 2021, doi: 10.1177/17543371211054752.
[231] J. Lee, K. Wheeler, and D. A. James, *Wearable Sensors in Sport: A Practical Guide to Usage and Implementation*, 1st ed. Singapore: Springer, 2019.
[232] K. Austin, K. J. Nicholas, C. M. Jones, and M. Loosemore, "Criterion validity and reliability of an instrumented mouthguard under pendulum impactor conditions," *Sports Engineering*, vol. 26, no. 1, p. 41, Dec. 2023, doi: 10.1007/s12283-023-00434-4.
[233] R. Likert, "A technique for the measurement of attitudes," *Archives of Psychology*, vol. 22, 1932.
[234] G. T. G. Hughes, V. Camomilla, B. Vanwanseele, A. J. Harrison, D. T. P. Fong, and E. J. Bradshaw, "Novel technology in sports biomechanics: some words of caution," *Sports Biomech*, vol. 23, no. 4, pp. 393–401, Apr. 2024, doi: 10.1080/14763141.2020.1869453.
[235] "ISO 8783:2015 – Alpine skis – Guidelines for conducting slope performance tests." Accessed: May 1, 2025. [Online]. Available: https://www.iso.org/standard/65972.html
[236] A. Benghanem, J. Audet, and A. Lussier-Desbiens, "Alpine skis categorization with comprehensive decision tree models," *Sports Engineering*, vol. 26, no. 1, Dec. 2023, doi: 10.1007/s12283-023-00405-9.
[237] D. Heinrich, M. Mössner, P. Kaps, H. Schretter, and W. Nachbauer, "Influence of ski bending stiffness on the turning radius of alpine skis at different edging angles and velocities," in *The Engineering of Sport 6*, E. F. Moritz and S. Haake, Eds, New York: Springer, 2006, pp. 207–212. doi: 10.1007/978-0-387-46051-2_37.
[238] P. Federolf, M. Auer, M. Fauve, A. Lüthi, and H. Rhyner, "Subjective evaluation of the performance of alpine skis and correlations with mechanical ski properties," in *The Engineering of Sport 6*, E. F. Moritz and S. Haake, Eds, New York: Springer, 2006, pp. 287–292. doi: 10.1007/978-0-387-46050-5_51.

[239] C. Brousseau and A. L. Desbiens, "Alpine skiing recommendation tool and performance prediction," in *The 12th Conference of the International Sports Engineering Association*, H. G. Espinosa, D. R. Rowlands, J. Shepherd, and D. V. Thiel, Eds, Basel: MDPI, Mar. 2018, p. 316. doi: 10.3390/proceedings2060316.
[240] J. Truong, P. Gosselin, and A. L. Desbiens, "The effect of bending and torsional stiffness on the edge grip of an alpine ski," in *The Engineering of Sport 14*, T. Allen, H. Espinosa, and S. Shade, Eds, West Lafayette, IN: Purdue University, Jun. 2022, p. 6. doi: 10.5703/1288284317481.
[241] T. Payne, S. Mitchell, R. Bibb, and M. Waters, "Development of novel synthetic muscle tissues for sports impact surrogates," *J Mech Behav Biomed Mater*, vol. 41, pp. 357–374, Jan. 2015, doi: 10.1016/J.JMBBM.2014.08.011.
[242] "F1447 Standard Specification for Helmets Used in Recreational Bicycling or Roller Skating." Accessed: Jul. 21, 2025. [Online]. Available: https://store.astm.org/f1447-18.html
[243] "ISO 5725-1:2023 – Accuracy (trueness and precision) of measurement methods and results – Part 1: General principles and definitions." Accessed: May 1, 2025. [Online]. Available: https://www.iso.org/standard/69418.html
[244] A. S. Perrotta, B. D. Day, A. J. Scott, and E. A. Gnatiuk, "Precision and reliability of the polar team pro computer software for analyzing heart rate variability," *Sports Engineering*, vol. 26, no. 1, p. 40, Dec. 2023, doi: 10.1007/s12283-023-00433-5.
[245] A. S. Perrotta *et al.*, "Concurrent validity and reliability of photoelectric and accelerometer technology for calculating vertical jump height in female athletes," *Sports Engineering*, vol. 26, no. 1, p. 48, Dec. 2023, doi: 10.1007/s12283-023-00440-6.
[246] S. L. Schmidt, Ed., *21st Century Sports*. Cham: Springer, 2024.
[247] S. Dixon, P. Fleming, I. James, and M. Carré, *The Science and Engineering of Sport Surfaces*, 1st ed. London: Routledge, 2015.
[248] S. Choppin, S. Albrecht, J. Spurr, and J. Capel-Davies, "The effect of ball wear on ball aerodynamics: an investigation using hawk-eye data," in *The 12th Conference of the International Sports Engineering Association*, H. G. Espinosa, D. R. Rowlands, J. Shepherd, and D. V. Thiel, Eds, Basel: MDPI, Mar. 2018, p. 265. doi: 10.3390/proceedings2060265.
[249] S. J. Haake, S. G. Chadwick, R. J. Dignall, S. Goodwill, and P. Rose, "Engineering tennis – slowing the game down," *Sports Engineering*, vol. 3, no. 2, pp. 131–143, May 2000, doi: 10.1046/j.1460-2687.2000.00040.x.
[250] T. Allen, S. Choppin, and D. Knudson, "A review of tennis racket performance parameters," *Sports Engineering*, vol. 19, no. 1, Mar. 2016, doi: 10.1007/s12283-014-0167-x.
[251] D. R. Bassett, C. R. Kyle, L. Passfield, J. P. Broker, and E. R. Burke, "Comparing cycling world hour records, 1967–1996: modeling with empirical data," *Med Sci Sports Exerc*, vol. 31, no. 11, p. 1665, Nov. 1999, doi: 10.1097/00005768-199911000-00025.
[252] S. Haake, "Material advantage?" *Physics World*, vol. 25, no. 7, pp. 26–30, Jul. 2012, doi: 10.1088/2058-7058/25/07/33.
[253] S. Haake, D. James, and L. Foster, "An improvement index to quantify the evolution of performance in field events," *J Sports Sci*, vol. 33, no. 3, pp. 255–267, Feb. 2015, doi: 10.1080/02640414.2014.938099.
[254] S. J. Haake, L. I. Foster, and D. M. James, "An improvement index to quantify the evolution of performance in running," *J Sports Sci*, vol. 32, no. 7, pp. 610–622, Apr. 2014, doi: 10.1080/02640414.2013.841974.
[255] R. Lukes, J. Hart, and S. Haake, "An analytical model for track cycling," *Proc Inst Mech Eng P J Sport Eng Technol*, vol. 226, no. 2, pp. 143–151, Jun. 2012, doi: 10.1177/1754337111433242.
[256] A. Friis Bach, J. Alexandersen, and C. B. Lundgaard, "A numerical design methodology for optimal pacing strategy in the individual time trial discipline of cycling," *Sports Engineering*, vol. 28, no. 1, pp. 1–11, Mar. 2025, doi: 10.1007/S12283-025-00493-9.
[257] S. Haake, S. Choppin, T. Allen, and S. Goodwill, "The evolution of the tennis racket and its effect on serve speed," in *Tennis Science and Technology 3*, S. Miller and J. Capel-Davies, Eds, London: International Tennis Federation, 2007, pp. 257–271.
[258] S. Willwacher, P. Mai, J. Helwig, M. Hipper, B. Utku, and J. Robbin, "Does advanced footwear technology improve track and road racing performance? An explorative analysis based on the 100 best yearly performances in the world between 2010 and 2022," *Sports Med Open*, vol. 10, no. 1, pp. 1–9, Dec. 2024, doi: 10.1186/s40798-024-00683-y.

[259] J. Guinness, D. Bhattacharya, J. Chen, M. Chen, and A. Loh, "An observational study of the effect of Nike Vaporfly shoes on marathon performance," *ArXiv*, doi: 10.48550/arXiv.2002.06105.

[260] F. G. Beltrami, G.-A. Baumann, K. Biedermann, and C. M. Spengler, "On the fallacy of single trials: a proposal to reduce variability in running shoe research," *Sports Medicine*, pp. 1–5, Mar. 2025, doi: 10.1007/S40279-025-02180-Y.

[261] Z. B. Barrons, V. Rodrigo-Carranza, M. Bertschy, and W. Hoogkamer, "The fallacy of single trials: the need for multiple trials in assessing running economy responses in advanced footwear technology," *Sports Medicine*, vol. 54, no. 6, pp. 1357–1360, Jun. 2024, doi: 10.1007/s40279-023-01991-1.

[262] "How 'shoe doping' changed marathon times forever – in ways we still don't fully understand – The Athletic." Accessed: Mar. 20, 2025. [Online]. Available: https://www.nytimes.com/athletic/5834953/2024/10/12/shoe-doping-marathon-times-kipchoge/

[263] S. Bermon, F. Garrandes, A. Szabo, I. Berkovics, and P. E. Adami, "Effect of advanced shoe technology on the evolution of road race times in male and female elite runners," *Front Sports Act Living*, vol. 3, p. 653173, Apr. 2021, doi: 10.3389/FSPOR.2021.653173.

[264] J. Mason, L. Starc, J. B. Morin, E. L. McClelland, and A. Zech, "Can the recent sex-specific evolutions in elite running performances be attributed to advanced footwear technology?" *Front Sports Act Living*, vol. 6, p. 1386627, May 2024, doi: 10.3389/FSPOR.2024.1386627.

[265] M. Knopp *et al.*, "Variability in running economy of Kenyan world-class and European amateur male runners with advanced footwear running technology: experimental and meta-analysis results," *Sports Medicine*, vol. 53, no. 6, pp. 1255–1271, Jun. 2023, doi: 10.1007/s40279-023-01816-1.

[266] V. Rodrigo-Carranza, F. González-Mohíno, J. Santos-Concejero, and J. M. González-Ravé, "Impact of advanced footwear technology on elite men's in the evolution of road race performance," *J Sports Sci*, vol. 40, no. 23, pp. 2661–2668, 2022, doi: 10.1080/02640414.2023.2183103.

[267] S. Choppin, "Calculating football drag profiles from simulated trajectories," *Sports Engineering*, vol. 16, no. 3, pp. 189–194, Sep. 2013, doi: 10.1007/s12283-013-0124-0.

[268] J. E. Goff, J. Kelley, C. M. Hobson, K. Seo, T. Asai, and S. B. Choppin, "Creating drag and lift curves from soccer trajectories," *Eur J Phys*, vol. 38, no. 4, p. 044003, May 2017, doi: 10.1007/s12283-013-0124-0.

[269] J. Truong, M. Bulota, and A. L. Desbiens, "Historical trends in alpine ski design: how skis have evolved over the past century," in *The 13th Conference of the International Sports Engineering Association*, M. Nakashima, T. Maruyama, and Y. Miyazaki, Eds, Basel: MDPI, Jun. 2020, p. 135. doi: 10.3390/proceedings2020049135.

[270] T. Eftaxiopoulou, L. Persad, and A. M. J. Bull, "Assessment of performance parameters of a series of five 'historical' cricket bat designs," *Proc Inst Mech Eng P J Sport Eng Technol*, vol. 231, no. 1, 2017, doi: 10.1177/1754337116638970.

[271] J. A. Teixeira da Silva *et al.*, "The video assistant referee in football," *Sports Engineering*, vol. 27, no. 1, pp. 1–15, Jun. 2024, doi: 10.1007/s12283-024-00459-3.

[272] Fédération Internationale de Football Association (FIFA), "Environmental protection in the production of footballs," Zurich, 2019. Accessed: May 4, 2025. [Online]. Available: https://inside.fifa.com/innovation/news/sustainability-meets-football-production

[273] C. H. M. Chan, M. K. I. Ma, T. K. Chan, L. T. Yeung, and Y. K. Chan, "Current applications of virtual reality in basketball training: a systematic review," *Sports Engineering*, vol. 27, no. 2, p. 26, Dec. 2024, doi: 10.1007/s12283-024-00469-1.

[274] K. Petri *et al.*, "Training using virtual reality improves response behavior in karate kumite," *Sports Engineering*, vol. 22, no. 1, p. 2, Mar. 2019, doi: 10.1007/s12283-019-0299-0.

[275] Y. Ritter, M. Droste, D. Bürger, S. Pastel, and K. Witte, "Comparison of response behavior in karate kumite between real world and virtual reality," *Sports Engineering*, vol. 25, no. 1, p. 14, Dec. 2022, doi: 10.1007/s12283-022-00378-1.

[276] D. Memmert, Ed., *Sports Technology: Technologies, Fields of Application, Sports Equipment and Materials for Sport*, 1st ed. Berlin: Springer, 2024.

[277] I. Fister, I. Fister Jr., and D. Fister, *Computational Intelligence in Sports*. Cham: Springer, 2019.

[278] Y. Hong, Ed., *Routledge Handbook of Ergonomics in Sport and Exercise*, 1st ed. London: Routledge, 2013.

[279] M. Peters, Ed., *Computational Fluid Dynamics for Sport Simulation*, 1st ed. Berlin: Springer, 2009.

[280] R. S. Goonetilleke, Ed., *The Science of Footwear*. New York: CRC Press, 2013.
[281] R. Minton, *Golf by the Numbers: How Stats, Math, and Physics Affect Your Game*. Baltimore, MD: Johns Hopkins University Press, 2012.
[282] L. Seifert, D. Chollet, and I. I. Mujika, *World Book of Swimming: From Science to Performance*. Hauppauge, NY: Nova Science Publishers Inc, 2011.
[283] V. Medved, *Measurement of Human Locomotion*, 1st ed. Boca Raton, FL: CRC Press, 2000.
[284] R. Cross and C. Lindsey, *Technical Tennis: Racquets, Strings, Balls, Courts, Spin, and Bounce*, 1st ed. Vista, CA: Racquet Tech Publishing, 2005.
[285] H. Brody, R. Cross, and C. Lindsey, *The Physics and Technology of Tennis*, 1st ed. Vista, CA: Racquet Tech Publishing, 2002.
[286] F. Braghin, F. Cheli, S. Melzi, E. Sabbioni, and S. Maldifassi, *The Engineering Approach to Winter Sports*, 1st ed. New York: Springer, 2016.

Index

Locators in **bold** refer to tables and those in *italics* to figures.

For Product Safety Concerns and Information please contact our EU representative GPSR@taylorandfrancis.com Taylor & Francis Verlag GmbH, Kaufingerstraße 24, 80331 München, Germany

Batch number: 10722134

Printed by Printforce, the Netherlands